# PARANORMAL

*A Critical Analysis*

Edited by
## Bryan Farha

**Foreword by Michael Shermer**

**University Press of America,® Inc.**
Lanham · Boulder · New York · Toronto · Plymouth, UK

Copyright © 2007 by
University Press of America,® Inc.
4501 Forbes Boulevard
Suite 200
Lanham, Maryland 20706
UPA Acquisitions Department (301) 459-3366

Estover Road
Plymouth PL6 7PY
United Kingdom

Library of Congress Control Number: 2007925508
ISBN-13: 978-0-7618-3772-5 (paperback : alk. paper)
ISBN-10: 0-7618-3772-8 (paperback : alk. paper)

⊖™ The paper used in this publication meets the minimum
requirements of American National Standard for Information
Sciences—Permanence of Paper for Printed Library Materials,
ANSI Z39.48—1984

*To the memory of Carl Sagan,*

*whose presentation of the universe in* Cosmos

*served as an inspiration for the rest*

*of my personal and professional life*

# CONTENTS

# FOREWORD

## SCIENCE AND PSEUDOSCIENCE: THE DIFFERENCE IN THINKING AND THE DIFFERENCE IT MAKES

### Michael Shermer

*Atheists abound in these days and witchcraft is called into question. If neither possession nor witchcraft (contrary to what has been so long generally and confidently affirmed), why should we think that there are devils? If no devils, no God.*

So wrote one observer in the seventeenth century, out of religious concern that atheism might ascend to social respectability along with science. The study of demons, witches, and spirits, in fact, took a decidedly empirical turn in the early modern period, along with many other knowledge traditions, with the goal of incorporating its rapidly growing respectability. As a consequence, the lines between science and nonscience grew blurry.

Since the rise of modern science in the sixteenth and seventeenth centuries, attempts to adjudicate the difference between science and other knowledge traditions have always been more than an exercise in academic debate. The religious, political, and social implications of how science is defined, who defines it, and who and what is left out of the definition has been a contentious one. Today, the term pseudoscience is often employed by those in the scientific community to disparage claims to scientific credibility that, in fact, lack evidence or fail to employ the methods of science.

### The Demarcation Problem

In the twentieth century the philosophy of science developed into a viable academic discipline, out of which grew attempts to delimit science and nonscience traditions. In *The Logic of Scientific Discovery*, for example, the philosopher of science Karl Popper identified what he called "the problem of demarcation," that is "the problem of finding a criterion which would enable us to distinguish between the empirical sciences on the one hand, and mathematics and logic as well as 'metaphysical' systems on the other" (Popper 1934, 27). Most scientists and philosophers use induction as the criterion of demarcation— if one reasons from particular observations or singular statements to universal theories or general conclusions, then one is doing empirical science. Popper's thesis was that induction does not actually provide empirical proof—"no matter how many instances of white swans we may have observed, this does not justify the conclusion that all swans are white" (34)—and that, de facto, scientists actually reason deductively, from the universal and general to the singular and particular. But in rejecting induction as the preferred (by others) criterion of demarcation between science and nonscience, Popper was concerned that his emphasis on deduction would lead to an inevitable fuzziness of the boundary line. If a scientific theory can never actually be proven, then is science no different from other knowledge disciplines?

Popper's solution to the problem of demarcation was the criterion of falsifiability. Theories are "never empirically verifiable," but if they are falsifiable then they belong in the domain of empirical science. "In other words: I shall not require of a scientific system that it shall be capable of being singled out, once and for all, in a positive sense; but I shall require that its logical form shall be such that it can be singled out, by means of empirical tests, in a negative sense: it must be possible for an empirical scientific system to be refuted by experience" (Popper 1934, 70). The theory of evolution, for example, has been accused by creationists as being nonscientific because no one was there to see it happen and biologists cannot observe it in the laboratory because it takes too long. But, in fact, by Popper's criterion of falsifiability, the theory of evolution would be doomed to the trash heap of bad science if, say, human fossil remains turned up in the same geological bedding planes as 300-million-year-old trilobites. No such falsification of evolution has ever been found, and although by Popper's criterion this does not mean that the theory has been proven absolutely, it does mean that it has yet to be falsified, thus placing it firmly in the camp of solid empirical science.

## Science Defended, Science Defined

The evolution-creationism controversy, in fact, has provided both scientific and legal forms of demarcation between science and pseudoscience. It is one thing for academic scientists and philosophers to debate the definition of science; it is another matter when the U.S. Supreme Court weighs in on the issue. Because evolution could not be excluded from public school science classrooms in the early twentieth century, and because the teaching of religious tenets was deemed unconstitutional in a number of state trials in the middle of the twentieth century, in the latter part of the century creationists began to call their doctrines creation-science. Since academic openness calls for a balanced treatment of competing ideas, they argued, creation-science should be taught side by side with evolution-science. In 1982 creationists succeeded in getting passed the Louisiana Balanced Treatment for Creation-Science and Evolution Science Act. In 1985 the law was struck down in the Federal Court of Louisiana, a decision that was appealed to the U.S. Court of Appeals for the Fifth Circuit. In 1986 the U.S. Supreme Court agreed to hear the case, leading to the publication of a remarkable document that clearly and succinctly adjudicated (literally in this case) the difference between science and pseudoscience.

The document was an amicus curiae brief submitted to the court on behalf of seventy-two Nobel laureates in science, seventeen state academies of science, and seven other scientific organizations. The amicus brief begins by offering a general definition, "Science is devoted to formulating and testing naturalistic explanations for natural phenomena. It is a process for systematically collecting and recording data about the physical world, then categorizing and studying the collected data in an effort to infer the principles of nature that best explain the observed phenomena." Next, the scientific method is discussed, beginning with the collection of "facts," the data of the world. "The grist for the mill of scientific inquiry is an ever increasing body of observations that give information about underlying 'facts.' Facts are the properties of natural phenomena. The scientific method involves the rigorous, methodical testing of principles that might present a naturalistic explanation for those facts" (1986, 23).

Based on well-established facts, testable hypotheses are formed. The process of testing "leads scientists to accord a special dignity to those hypotheses that accumulate substantial observational or experimental support." This "special dignity" is called a "theory" that, when it "explains a large and diverse body of facts" is considered "robust" and if it "consistently predicts new phenomena that are subsequently observed" it is "reliable." Facts and theories are not to be used interchangeably or in relation to one another as more or less true. Facts are the world's data. Theories are explanatory ideas about those facts. "An explanatory principle is not to be confused with the data it seeks to explain." Constructs and other nontestable statements are not a part of science. "An explanatory principle that by its nature cannot be tested is outside the realm of science" (23-24).

It follows from the nature of scientific method that no explanatory principles in science are final. "Even the most robust and reliable theory...is tentative. A scientific theory is forever subject to reexamination and—as in the case of Ptolemaic astronomy—may ultimately be rejected after centuries of viability." Scientists encounter uncertainty as a regular and natural part of their work. "In an ideal world, every science course would include repeated reminders that each theory presented to explain our observations of the universe carries this qualification: 'as far as we know now, from examining the evidence available to us today'" (1986, 24). Science also seeks only naturalistic explanations for phenomena. "Science is not equipped to evaluate supernatural explanations for our observations; without passing judgment on the truth or falsity of supernatural explanations, science leaves their consideration to the domain of religious faith" (23). According to the amicus any body of knowledge accumulated within the guidelines previously described is considered scientific and suitable for public school science education; and any body of knowledge not accumulated within these guidelines is not considered scientific.

On June 19, 1987, the U.S. Supreme Court, by a vote of 7 to 2, held that the Louisiana Act "is facially invalid as violative of the Establishment Clause of the First Amendment, because it lacks a clear secular purpose" and that "[t]he Act impermissibly endorses religion by advancing the religious belief that a supernatural being created humankind" (*Edwards v. Aguillard*, 1987, 1). The Louisiana trial in general, and the amicus brief in particular, had the effect of temporarily galvanizing the scientific community into defining science as a body of knowledge accumulated through a particular scientific method, as defined by the leading members of the scientific community themselves. Science is as scientists do.

## Delimiting the Boundaries Between Science and Pseudoscience

Creation-science (and its most recent hybrid, intelligent design theory) is just one of many claims that most mainstream scientists reject as pseudoscience. But what about those claims to scientific knowledge that are not so obviously classified as pseudoscience? When encountering a claim, how can one determine whether it constitutes a legitimate assertion as scientific? What follows are a list of ten questions that get to the heart of delimiting the boundaries between science and pseudoscience:

*(1) How reliable is the source of the claim?* All scientists make mistakes, but are the mistakes random, as one might expect from a normally reliable source, or are they directed toward supporting the claimant's preferred belief? Scientists' mistakes tend to be random; pseudoscientists' mistakes tend to be directional.

*(2) Does this source often make similar claims?* Pseudoscientists have a habit of going well beyond the facts, and so when individuals make many extraordinary claims, they may be more than iconoclasts. What one is looking for here is a pattern of fringe thinking that consistently ignores or distorts data.

*(3) Have the claims been verified by another source?* Typically, pseudoscientists make statements that are unverified or are verified by a source within their own belief circle. One must ask who is checking the claims and even who is checking the checkers.

*(4) How does the claim fit with what is known about how the world works?* An extraordinary claim must be placed in a larger context to see how it fits. When people claim that the pyramids and the Sphinx were built over 10,000

years ago by an advanced race of humans, they are not presenting any context for that earlier civilization. Where are its works of art, weapons, clothing, tools, and trash?

*(5) Has anyone made an effort to disprove the claim or has only confirmatory evidence been sought?* This is the confirmation bias or the tendency to seek confirmatory evidence and reject or ignore disconfirmatory evidence. The confirmation bias is powerful and pervasive. This is why the scientific method, which emphasizes checking and rechecking, verification and replication, and especially attempts to falsify a claim, is critical.

*(6) Does the preponderance of evidence converge on the claimant's conclusion or a different one?* The theory of evolution, for example, is proved through a convergence of evidence from a number of independent lines of inquiry. No single fossil or piece of biological or paleontological evidence has the word evolution written on it; instead, there is a convergence from tens of thousands of evidentiary bits that adds up to a story of the evolution of life. Creationists conveniently ignore this convergence, focusing instead on trivial anomalies or currently unexplained phenomena in the history of life.

*(7) Is the claimant employing the accepted rules of reason and tools of research or have those rules and tools been abandoned in favor of others that lead to the desired conclusion?* UFOlogists exhibit this fallacy in their continued focus on a handful of unexplained atmospheric anomalies and visual misperceptions by eyewitnesses while ignoring that the vast majority of UFO sightings are fully explicable.

*(8) Has the claimant provided a different explanation for the observed phenomena or is it strictly a matter of denying the existing explanation?* This is a classic debate strategy: Criticize your opponent and never affirm what you believe to avoid criticism. This strategy is unacceptable in science.

*(9) If the claimant has proffered a new explanation, does it account for as many phenomena as does the old explanation?* For a new theory to displace an old theory, it must explain what the old theory did and then some.

*(10) Do the claimants' personal beliefs and biases drive the conclusions or vice versa?* All scientists have social, political, and ideological beliefs that potentially could slant their interpretations of the data, but at some point, usually during the peer-review system, those biases and beliefs are rooted out or the paper or book is rejected for publication.

## The Enchanted Glass of Science

At the dawn of the scientific revolution in the early seventeenth century, the English philosopher Francis Bacon sought to turn away from the scholastic tradition of logic and reason as the sole road to truth, as well as reject the Renaissance quest to restore the perfection of ancient Greek knowledge. In his 1620 work *Novum Organum* (*New Tool*, contrary to the opinion of Aristotle's

*Organon*), Bacon portrayed science as humanity's savior that would inaugurate a restoration of all natural knowledge through a proper blend of observation and logic, data and theory. Bacon understood, however, that there are significant social and psychological barriers that interfere with one's understanding of the natural world, "For the mind of man is far from the nature of a clear and equal glass, wherein the beams of things should reflect according to their true incidence; nay, it is rather like an enchanted glass, full of superstition and imposture, if it be not delivered and reduced" (Bacon 1620, 53). In the end, thought Bacon, science offers the best hope to deliver the mind from such superstition and imposture. Today, science continues to deliver on that hope.

## BIBLIOGRAPHY

Amicus Curiae Brief of 72 Nobel Laureates, 17 State Academics of Science, and 7 Other Scientific Organizations, in Support of Appellees, Submitted to the Supreme Court of the United States, October Term, 1986, as Edwin W. Edwards, in His Official Capacity as Governor of Louisiana, et al., Appellants v. Don Aguillard et al., Appellees, 1986.

Bacon, Francis. *Francis Bacon: A Selection of His Works.* Edited by Sidney Warhaft. Toronto: Macmillan, 1965.

Popper, Karl. *The Logic of Scientific Discovery.* New York: Harper and Row, 1934.

# PREFACE

*I believe that the extraordinary should be pursued. But extraordinary claims require extraordinary evidence.*

Carl Sagan

*Paranormal Claims: A Critical Analysis* is an academic reader of invited articles regarding extraordinary, unexplained, or pseudoscientific phenomena. From a rational, logical, and skeptical perspective, the topics examined include astrology, psychic ability, alternative medicine, UFOs and aliens, afterlife communication, Bigfoot, numerical anomalies, and faith healing. These articles were selected because of their quality, importance, and topical visibility. The contributors are among the most accomplished critical thinkers, scientists, and educators in the world. Indeed, the contributor list reads like a "Who's Who" of skeptical scientists, educators, and thinkers. Many of these articles are classics. Some publication dates may seem outdated, but each article is a "must read" for those interested in skepticism. I have been in higher education for over twenty years, and I seldom find students who are excited to study "critical thinking." That changes when the subject matter is allegedly paranormal phenomena. This book, then, is a golden opportunity for the student or general reader interested in critical thought. Educators will notice increased levels of engagement from students.

## Why Bother?

I'm often asked why I bother combating unsubstantiated extraordinary claims and pseudoscience. What's the harm? What's the danger? It may be relatively harmless to be told by a carnival psychic that as a child your favorite color was blue and that you liked cheeseburgers. Beyond the entertainment of a sideshow, however, there is much more at stake. Many of today's psychics are con artists with the intent to siphon away our money—or worse—using clearly deceptive practices. A psychic claiming to diagnose health conditions can be dangerous. For example, if someone gets a "reading" and is told by that psychic that an upset stomach is caused by stomach cancer—when it isn't—he or she may spend hundreds or thousands of dollars in unnecessary medical attention. And he or she has needlessly worried about dying. On the other hand, if a psychic tells someone in the early stages of stomach cancer that he or she "sees" simple gastritis, the client may delay necessary medical attention until it is too late for conventional, possibly lifesaving   treatment. In the absence of evidence to support their claims, then, psychics who perform health readings are potentially dangerous.

Consider issues such as these if someone asks why you would waste time reading a book on the paranormal. And tell them it's not exactly a book about the paranormal—it's a book about thinking critically while using alleged paranormal phenomena for a subject. *But the book is about critical thinking.*

## Critical Thinking

How can we learn to think critically? If an astrologer explains a certain behavior by claiming that "the planet Mercury is in retrograde," we might ask what "retrograde" motion is. Check it out to see if it makes sense. Many astrologers (pseudoscientists) think retrograde is a reverse orbit around the Sun, which would be extremely peculiar. But astronomers (scientists) tell us that retrograde is an *apparent* backward motion—in other words, an illusion. Regardless of what astrologers claim, *all planets orbit the Sun in the same direction*, so how could apparent backward motion—a visual effect—influence behavior? This is a critical thinking question. But what if we don't care about astrology—how can this information still be useful? The answer is that the specific information may not be useful, but the skill of learning how to think critically is. The more we learn how to think critically, the better we'll be able to function in society and make wise, informed decisions. Thinking critically can help us to avoid being swindled by scam artists, find a more suitable job, or select better schools for our children. The benefits are endless. Thus, critical thinking is helpful in ways far beyond merely teaching us how psychics can fool people into believing they have special powers. This book is an excellent exercise in critical thinking.

Critical thinking also guards against jumping to conclusions or "assuming." For instance, just because I've included articles in this book that are skeptical of afterlife communication and faith healing doesn't mean that I'm closed to the concept of an afterlife or divine creation. But that, too, doesn't mean that I believe Sylvia Browne can communicate with the dead or that Benny Hinn can reverse illness via a direct pipeline to God.

I often wonder why some of the most famous psychics in the world have never taken the One Million Dollar Paranormal Challenge offered by the James Randi Educational Foundation. The challenge offers prize money to anyone who can demonstrate psychic ability under controlled testing conditions. Psychics have many excuses for not taking the challenge, including statements that they are not motivated by money. It is refreshing to hear that someone isn't driven by money, of course, but why couldn't the psychic take the challenge, win the prize money, and donate it to the charity of their choice? And, if they're not motivated by money, why aren't they using their "gifts" to help locate the most heinous terrorists in the world? Again—*critical thinking*.

There's usually nothing extraordinary about selling a house. A prospective buyer, though, will scrutinize the house from top to bottom, inside and out. He or she will look over the paint, go in the attic, check out the closets, and have a professional inspection if serious about making the purchase. But why is it that the same person who so heavily scritinizes the house will almost blindly believe in the extraordinary claims of a psychic, astrologer, homeopathic practitioner, or ghost hunter? Why do some people seem to abandon critical thinking when it comes to the paranormal?

## *Skeptical Misconceptions*

There are a few misconceptions regarding people who are skeptical that needs some commentary. First is the assumption that we are completely close-minded. This is simply untrue because to be skeptical is to *question*—not to *disbelieve*. But those of us with an appreciation for the scientific method realize the importance of utilizing control measures in experimentation. So what does this have to do with the issue of being close-minded? Adhering to control measures is vital to science and it can often mislead the public into believing that skeptical researchers are consciously devising experiments that cannot be passed by a claimant—thereby giving the impression that we disbelieve everything extraordinary. What the experimenter is actually doing is devising a test that can't be passed *unless the claimant can perform the extraordinary feat that is alleged*. That's just good science—not close-mindedness. Having said that, I'd like to turn the tables by challenging the non-skeptical reader to approach this book with an open mind. A student in one of my classes recently mentioned that she's a "believer" in psychics because science hasn't proven its impossibility— to which I replied "but if it's empirically testable, and experiments to date don't support psychic claims, then why would you draw the conclusion of its reality?" Where testable, I hope that beliefs are formulated and conclusions are drawn based on available *evidence*—not feelings or impressions.

A related misconception is that we not only are disbelieving, but that we are *automatically* "disbelieving" in just about everything. This is also untrue of a good skeptic. According to *Guinness World Records*, Robert Pershing Wadlow (1918–1940) is still the tallest human in recorded history at 8 feet, 11 inches in height. Obviously, then, we know people *can* grow that tall. So does that mean that someone should automatically believe me if I tell them I have a 9 foot brother? Should they be automatically disbelieving? It's possible to grow that tall, right? If someone can grow to 8 feet, 11 inches, why can't a person grow to 9 feet? But it's such an extraordinary claim that, understandably, I should expect some scrutiny. I might be asked to provide a photograph or to present my 9 foot brother in person. Shouldn't the scrutiny be acceptable, though, if I'm telling the truth? If I'm not telling the truth, then I will likely live with the deception and maintain the lie forever. Do you think this happens? But many claims such as extra tall humans, talking to the dead, predicting behavior based on planetary positions at birth, and levitation *are* testable. So let's test and arrive at a definitive conclusion where possible. We question—rather than disbelieve—until evidence avails itself sufficient enough to draw a responsible conclusion.

## *Reasons for Believing*

The truth is that people believe in allegedly paranormal phenomena, in part, because they have a *need* to believe—especially if it brings a feeling of comfort. For example, I've often been asked why an intelligent person would believe in ghosts. The answer is two-fold. First, Michael Shermer put it best when he said, "Smart people believe weird things because they are skilled at defending beliefs they arrived at for non-smart reasons." I love that statement. Second, the reason that some may believe in ghosts is the same reason they believe in near-death experiences or haunted houses—because belief in any of those things *necessarily* implies an afterlife. And the thought of an afterlife brings a comforting feeling, right? I'm not automatically disbelieving of an afterlife, though—because we don't know if one exists. We have to wait until death to find out. But since we can't set up an experiment to determine if an afterlife exists, my mind is open to the possibility. But the *need* isn't there to comfort me now. An afterlife—especially one that brings eternal bliss—would be a nice thing. But I don't want my beliefs to be determined by hopes, wants, needs, dreams, or desires. The universe is the way it is—not the way I *wish* it to be.

## *The Answer*

Did you know that there are more stars than there are grains of sand on all the beaches in the world? Did you know that oysters can change from one gender to another and back again? Did you know that on the planet Venus a day lasts longer than a year? Did you know the rings of Saturn span more than two and a half times the distance between the Earth and the Moon? Did you know that an elephant trunk has about 40,000 muscles? Did you know that ocean waves can travel as fast as a jet plane? Did you know there are more atoms in this letter "i" than there are seconds in ten thousand million years? If you want to study the extraordinary, study *science*. That is where the truly amazing resides, and it's absolutely real.

Bryan Farha  
Oklahoma City, Oklahoma  
January, 2007

# ACKNOWLEDGMENTS

I would like to express my gratitude to the late Carl Sagan for inspiring me to develop and promote critical thinking. Special thanks also goes to Michael Shermer, publisher and editor-in-chief of *Skeptic* magazine, for writing the foreword. Particular appreciation for endorsing the book with blurbs goes to Ann Druyan, co-writer of the Emmy and Peabody award-winning television series *Cosmos*; James Randi; and Ken Frazier, editor of *Skeptical Inquirer*. Their reputations speak for themselves, and their support is especially meaningful to me.

Additionally, I'd like to thank the following people for their technical consultation: Susan Barber, Terry Conley, and John Nail of Oklahoma City University; Claus Larsen of Denmark; Alice Stanton and Jay Dew of Norman, Oklahoma; Dayna Dempsey of Oklahoma City; and Mark Samara of Los Angeles.

I would like to extend my gratitude to the following people and publishers for allowing me to reprint materials: Ann Druyan, the estate of Carl Sagan, and Democritus Properties; Michael Shermer of *Skeptic* magazine; Barry Karr of *Skeptical Inquirer*; Stephen Barrett of Quackwatch.org and the National Council Against Health Fraud; James Randi of the James Randi Educational Foundation; Barry Beyerstein of Simon Fraser University; and Geoffrey Dean of Perth, Western Australia.

# 1

# THE FINE ART OF BALONEY DETECTION

## Carl Sagan

My parents died years ago. I was very close to them. I still miss them terribly. I know I always will. I long to believe that their essence, their personalities, what I loved so much about them, are—really and truly—still in existence somewhere. I wouldn't ask very much, just five or ten minutes a year, say, to tell them about their grandchildren, to catch them up on the latest news, to remind them that I love them. There's a part of me—no matter how childish it sounds—that wonders how they are. "Is everything all right?" I want to ask. The last words I found myself saying to my father, at the moment of his death, were "Take care."

Sometimes I dream that I'm talking to my parents, and suddenly—still immersed in the dreamwork—I'm seized by the overpowering realization that they didn't really die, that it's all been some kind of horrible mistake. Why, here they are, alive and well, my father making wry jokes, my mother earnestly advising me to wear a muffler because the weather is chilly. When I wake up I go through an abbreviated process of mourning all over again. Plainly, there's something within me that's ready to believe in life after death. And it's not the least bit interested in whether there's any sober evidence for it.

So I don't guffaw at the woman who visits her husband's grave and chats him up every now and then, maybe on the anniversary of his death. It's not hard to understand. And if I have difficulties with the ontological status of who she's talking to, that's all right. That's not what this is about. This is about humans being human. More than a third of American adults believe that on some level they've made contact with the dead. The number seems to have jumped by 15 percent between 1977 and 1988. A quarter of Americans believe in reincarnation.

But that doesn't mean I'd be willing to accept the pretensions of a "medium," who claims to channel the spirits of the dear departed, when I'm aware the practice is rife with fraud. I know how much I want to believe that my parents have just abandoned the husks of their bodies, like insects or snakes molting, and gone somewhere else. I understand that those very feelings might make me easy prey even for an unclever con, or for normal people unfamiliar with their unconscious minds, or for those suffering from a dissociative psychiatric disorder. Reluctantly, I rouse some reserves of skepticism.

How is it, I ask myself, that channelers never give us verifiable information otherwise unavailable? Why does Alexander the Great never tell us about the exact location of his tomb, Fermat about his Last Theorem, John Wilkes Booth about the Lincoln assassination conspiracy, Hermann Goring about the Reichstag fire? Why don't Sophocles, Democritus, and Aristarchus dictate their lost books? Don't they wish future generations to have access to their masterpieces?

If some good evidence for life after death were announced, I'd be eager to examine it; but it would have to be real scientific data, not mere anecdote. As with the face on Mars and alien abductions, better the hard truth, I say, than the comforting fantasy. And in the final tolling it often turns out that the facts are more comforting than the fantasy.

The fundamental premise of "channeling," spiritualism, and other forms of necromancy is that when we die we don't. Not exactly. Some thinking, feeling, and remembering part of us continues. That whatever-it-is—a soul or spirit, neither matter nor energy, but something else—can, we are told, re-enter the bodies of human and other beings in the future, and so death loses much of its sting. What's more, we have an opportunity, if the spiritualist or channeling contentions are true, to make contact with loved ones who have died.

J. Z. Knight of the State of Washington claims to be in touch with a 35,000-year-old somebody called "Ramtha." He speaks English very well, using Knight's tongue, lips and vocal chords, producing what sounds to me to be an accent from the Indian Raj. Since most people know how to talk, and many—from children to professional actors—have a repertoire of voices at their command, the simplest hypothesis is that Ms. Knight makes "Ramtha" speak all by herself, and that she has no contact with disembodied entities from the Pleistocene Ice Age. If there's evidence to the contrary, I'd love to hear it. It would be considerably more impressive if Ramtha could speak by himself, without the assistance of Ms. Knight's mouth. Failing that, how might we test the claim? (The actress Shirley MacLaine attests that Ramtha was her brother in Atlantis, but that's another story.)

Suppose Ramtha were available for questioning. Could we verify whether he is who he says he is? How does he know that he lived 35,000 years ago, even approximately? What calendar does he employ? Who is keeping track of the intervening millennia? Thirty-five thousand plus or minus what? What were things like 35,000 years ago? Either Ramtha really is 35,000 years old, in which case we discover something about that period, or he's a phony and he'll (or rather she'll) slip up.

Where did Ramtha live? (I know he speaks English with an Indian accent, but where 35,000 years ago did they do that?) What was the climate? What did

Ramtha eat? (Archaeologists know something about what people ate back then.) What were the indigenous languages, and social structure? Who else did Ramtha live with—wife, wives, children, grandchildren? What was the life cycle, the infant mortality rate, the life expectancy? Did they have birth control? What clothes did they wear? How were the clothes manufactured? What were the most dangerous predators? Hunting and fishing implements and strategies? Weapons? Endemic sexism? Xenophobia and ethnocentrism? And if Ramtha came from the "high civilization" of Atlantis, where are the linguistic, technological, historical and other details? What was their writing like? Tell us. Instead, all we are offered are banal homilies.

Here, to take another example, is a set of information channeled not from an ancient dead person, but from unknown non-human entities who make crop circles, as recorded by the journalist Jim Schnabel:

> We are so anxious at this sinful nation spreading lies about us. We do not come in machines, we do not land on your earth in machines . . . We come like the wind. We are Life Force. Life Force from the ground . . . Come here . . . We are but a breath away . . . a breath away . . . we are not a million miles away . . . a Life Force that is larger than the energies in your body. But we meet at a higher level of life . . . We need no name. We are parallel to your world, alongside your world . . . The walls are broken. Two men will rise from the past . . . the great bear . . . the world will be at peace.

People pay attention to these puerile marvels mainly because they promise something like old-time religion, but especially life after death, even life eternal.

A very different prospect for something like eternal life was once proposed by the versatile British scientist J. B. S. Haldane, who was, among many other things, one of the founders of population genetics. Haldane imagined a far future when the stars have darkened and space is mainly filled with a cold, thin gas. Nevertheless, if we wait long enough statistical fluctuations in the density of this gas will occur. Over immense periods of time the fluctuations will be sufficient to reconstitute a Universe something like our own. If the Universe is infinitely old, there will be an infinite number of such reconstitutions, Haldane pointed out.

So in an infinitely old universe with an infinite number of appearances of galaxies, stars, planets, and life, an identical Earth must reappear on which you and all your loved ones will be reunited. I'll be able to see my parents again and introduce them to the grandchildren they never knew. And all this will happen not once, but an infinite number of times.

Somehow, though, this does not quite offer the consolations of religion. If none of us is to have any recollection of what happened *this* time around, the time the reader and I are sharing, the satisfactions of bodily resurrection, in my ears at least, ring hollow.

But in this reflection I have underestimated what infinity means. In Haldane's picture, there will be universes, indeed an infinite number of them, in which our brains will have full recollection of many previous rounds. Satisfaction is at hand—tempered, though, by the thought of all those other universes which will also come into existence (again, not once but an infinite number of times) with tragedies and horrors vastly outstripping anything I've experienced this turn.

The Consolation of Haldane depends, though, on what kind of universe we live in, and maybe on such arcana as whether there's enough matter to eventually reverse the expansion of the universe, and the character of vacuum fluctuations. Those with a deep longing for life after death might, it seems, devote themselves to cosmology, quantum gravity, elementary particle physics, and transfinite arithmetic.

*     *     *

Clement of Alexandria, a Father of the early Church, in his *Exhortations to the Greeks* (written around the year 190) dismissed pagan beliefs in words that might today seem a little ironic:

> Far indeed are we from allowing grown men to listen to such tales. Even to our own children, when they are crying their heart out, as the saying goes, we are not in the habit of telling fabulous stories to soothe them.

In our time we have less severe standards. We tell children about Santa Claus, the Easter Bunny, and the Tooth Fairy for reasons we think emotionally sound, but then disabuse them of these myths before they're grown. Why retract? Because their well-being as adults depends on them knowing the world as it really is. We worry, and for good reason, about adults who still believe in Santa Claus.

On doctrinaire religions, "Men dare not avow, even to their own hearts," wrote the philosopher David Hume,

> the doubts which they entertain on such subjects. They make a merit of implicit faith; and disguise to themselves their real infidelity, by the strongest asseverations and the most positive bigotry.

This infidelity has profound moral consequences, as the American revolutionary Tom Paine wrote in *The Age of Reason:*

> Infidelity does not consist in believing, or in disbelieving; it consists in professing to believe what one does not believe. It is impossible to calculate the moral mischief, if I may so express it, that mental lying has produced in society. When man has so far corrupted and prostituted the chastity of his mind, as to subscribe his professional belief to things he does not believe, he has prepared himself for the commission of every other crime.

T. H. Huxley's formulation was

> The foundation of morality is to . . . give up pretending to believe that for which there is no evidence, and repeating unintelligible propositions about things beyond the possibilities of knowledge.

Clement, Hume, Paine, and Huxley were all talking about religion. But much of what they wrote has more general applications—for example to the pervasive background importunings of our commercial civilization: There is a class of aspirin commercials in which actors pretending to be doctors reveal the competing product to have only so much of the painkilling ingredient that doctors recommend most—they don't tell you what the mysterious ingredient is. Whereas *their* product has a dramatically larger amount (1.2 to 2 times more per tablet). So buy their product. But why not just take two of the competing tablets? Or consider the analgesic that works better than the "regular-strength" product of the competition. Why not then take the "extra-strength" competitive product? And of course they do not tell us of the more than a thousand deaths each year in the United States from the use of aspirin, or the roughly 5000 annual cases of kidney failure from the use of acetaminophen, chiefly Tylenol. Or who cares which breakfast cereal has more vitamins when we can take a vitamin pill with breakfast? Likewise, why should it matter whether an antacid contains calcium if the calcium is for nutrition and irrelevant for gastritis? Commercial culture is full of similar misdirections and evasions at the expense of the consumer. You're not supposed to ask. Don't think. Buy.

Paid product endorsements, especially by real or purported experts, constitute a steady rainfall of deception. They betray contempt for the intelligence of their customers. They introduce an insidious corruption of popular attitudes about scientific objectivity. Today there are even commercials in which real scientists, some of considerable distinction, shill for corporations. They teach that scientists too will lie for money. As Tom Paine warned, inuring us to lies lays the groundwork for many other evils.

I have in front of me as I write the program of one of the annual Whole Life Expos, New Age expositions held in San Francisco. Typically, tens of thousands of people attend. Highly questionable experts tout highly questionable products. Here are some of the presentations: "How Trapped Blood Proteins Produce Pain and Suffering." "Crystals, Are They Talismans or Stones?" (I have an opinion myself.) It continues: "As a crystal focuses sound and light waves for radio and television"—this is a vapid misunderstanding of how radio and television work—"so may it amplify spiritual vibrations for the attuned human." Or here's one "Return of the Goddess, a Presentational Ritual." Another: "Synchronicity, the Recognition Experience." That one is given by "Brother Charles." Or, on the next page, "You, Saint-Germain, and Healing Through the Violet Flame." It goes on and on, with plenty of ads about "opportunities"—running the short gamut from the dubious to the spurious—that are available at the Whole Life Expo.

Distraught cancer victims make pilgrimages to the Philippines, where "psychic surgeons," having palmed bits of chicken liver or goat heart, pretend to reach into the patient's innards and withdraw the diseased tissue, which is then triumphantly displayed. Leaders of Western democracies regularly consult astrologers and mystics before making decisions of state. Under public pressure for results, police with an unsolved murder or a missing body on their hands consult ESP "experts" (who never guess better than expected by common sense, but the police, the ESPers say, keep calling). A clairvoyance gap with adversary nations is announced, and the Central Intelligence Agency, under Congressional prodding, spends tax money to find out whether submarines in the ocean depths can be located by thinking hard at them. A "psychic"—using pendulums over maps and dowsing rods in airplanes—purports to find new mineral deposits; an Australian mining company pays him top dollar up front, none of it returnable in the event of failure, and a share in the exploitation of ores in the event of success. Nothing is discovered. Statues of Jesus or murals of Mary are spotted with moisture, and thousands of kind-hearted people convince themselves that they have witnessed a miracle.

These are all cases of proved or presumptive baloney. A deception arises, sometimes innocently but collaboratively, sometimes with cynical premeditation. Usually the victim is caught up in a powerful emotion—wonder, fear, greed, grief. Credulous acceptance of baloney can cost you money; that's what P. T. Barnum meant when he said, "There's a sucker born every minute." But it can be much more dangerous than that, and when governments and societies lose the capacity for critical thinking, the results can be catastrophic—however sympathetic we may be to those who have bought the baloney.

In science we may start with experimental results, data, observations, measurements, "facts." We invent, if we can, a rich array of possible explanations and systematically confront each explanation with the facts. In the course of their training, scientists are equipped with a baloney detection kit. The kit is brought out as a matter of course whenever new ideas are offered for consideration. If the new idea survives examination by the tools in our kit, we grant it warm, although tentative, acceptance. If you're so inclined, if you don't want to buy baloney even when it's reassuring to do so, there are precautions that can be taken; there's a tried-and-true, consumer-tested method.

What's in the kit? Tools for skeptical thinking.

What skeptical thinking boils down to is the means to construct, and to understand, a reasoned argument and—especially important—to recognize a fallacious or fraudulent argument. The question is not whether we like the conclusion that emerges out of a train of reasoning, but whether the conclusion follows from the premise or starting point and whether that premise is true.

Among the tools:

- **Wherever possible there must be independent confirmation of the "facts."**

- **Encourage substantive debate on the evidence** by knowledgeable proponents of all points of view.
- **Arguments from authority carry little weight**—"authorities" have made mistakes in the past. They will do so again in the future. Perhaps a better way to say it is that in science there are no authorities; at most, there are experts.
- **Spin more than one hypothesis.** If there's something to be explained, think of all the different ways in which it could be explained. Then think of tests by which you might systematically disprove each of the alternatives. What survives, the hypothesis that resists disproof in this Darwinian selection among "multiple working hypotheses," has a much better chance of being the right answer than if you had simply run with the first idea that caught your fancy.[1]
- **Try not to get overly attached to a hypothesis just because it's yours.** It's only a way station in the pursuit of knowledge. Ask yourself why you like the idea. Compare it fairly with the alternatives. See if you can find reasons for rejecting it. If you don't, others will.
- **Quantify.** If whatever it is you're explaining has some measure, some numerical quantity attached to it, you'll be much better able to discriminate among competing hypotheses. What is vague and qualitative is open to many explanations. Of course there are truths to be sought in the many qualitative issues we are obliged to confront, but finding *them* is more challenging.
- **If there's a chain of argument, *every* link in the chain must work** (including the premise)—not just most of them.
- **Occam's Razor.** This convenient rule-of-thumb urges us when faced with two hypotheses that explain the data *equally well* to choose the simpler.
- **Always ask whether the hypothesis can be, at least in principle, falsified.** Propositions that are untestable, unfalsifiable are not worth much. Consider the grand idea that our Universe and everything in it is just an elementary particle—an electron, say—in a much bigger Cosmos. But if we can never acquire information from outside our Universe, is not the idea incapable of disproof? You must be able to check assertions out. Inveterate skeptics must be given the chance to follow your reasoning, to duplicate your experiments and see if they get the same result.

The reliance on carefully designed and controlled experiments is key, as I tried to stress earlier. We will not learn much from mere contemplation. It is tempting to rest content with the first candidate explanation we can think of. One is much better than none. But what happens if we can invent several? How do we decide among them? We don't. We let experiment do it. Francis Bacon provided the classic reason:

Argumentation cannot suffice for the discovery of new work, since the subtlety of Nature is greater many times than the subtlety of argument.

Control experiments are essential. If, for example, a new medicine is alleged to cure a disease 20 percent of the time, we must make sure that a control population, taking a dummy sugar pill which as far as the subjects know might be the new drug, does not also experience spontaneous remission of the disease 20 percent of the time.

Variables must be separated. Suppose you're seasick, and given both an acupressure bracelet and 50 milligrams of meclizine. You find the unpleasantness vanishes. What did it—the bracelet or the pill? You can tell only if you take the one without the other, next time you're seasick. Now imagine that you're not so dedicated to science as to be willing to be seasick. Then you won't separate the variables. You'll take both remedies again. You've achieved the desired practical result; further knowledge, you might say, is not worth the discomfort of attaining it.

Often the experiment must be done "double-blind," so that those hoping for a certain finding are not in the potentially compromising position of evaluating the results. In testing a new medicine, for example, you might want the physicians who determine which patients' symptoms are relieved not to know which patients have been given the new drug. The knowledge might influence their decision, even if only unconsciously. Instead the list of those who experienced remission of symptoms can be compared with the list of those who got the new drug, each independently ascertained. Then you can determine what correlation exists. Or in conducting a police lineup or photo identification, the officer in charge should not know who the prime suspect is, so as not consciously or unconsciously to influence the witness.

\*          \*          \*

In addition to teaching us what to do when evaluating a claim to knowledge, any good baloney detection kit must also teach us what *not* to do. It helps us recognize the most common and perilous fallacies of logic and rhetoric. Many good examples can be found in religion and politics, because their practitioners are so often obliged to justify two contradictory propositions. Among these fallacies are:

- *ad hominem*—Latin for "to the man," attacking the arguer and not the argument (e.g., *The Reverend Dr. Smith is a known Biblical fundamentalist, so her objections to evolution need not be taken seriously*);
- **argument from authority** (e.g., *President Richard Nixon should be re-elected because he has a secret plan to end the war in Southeast Asia—* but because it was secret, there was no way for the electorate to

evaluate it on its merits; the argument amounted to trusting him because he was President: a mistake, as it turned out);

- **argument from adverse consequences** (e.g., *A God meting out punishment and reward must exist, because if He didn't, society would be much more lawless and dangerous—perhaps even ungovernable.*[2] Or: *The defendant in a widely publicized murder trial must be found guilty; otherwise, it will be an encouragement for other men to murder their wives*);

- **appeal to ignorance**—the claim that whatever has not been proved false must be true, and vice versa (e.g., *There is no compelling evidence that UFOs are not visiting the Earth; therefore UFOs exist—and there is intelligent life elsewhere in the Universe. Or: There may be seventy kazillion other worlds, but not one is known to have the moral advancement of the Earth, so we're still central to the Universe.*) This impatience with ambiguity can be criticized in the phrase: absence of evidence is not evidence of absence.

- **special pleading**, often to rescue a proposition in deep rhetorical trouble (e.g., *How can a merciful God condemn future generations to torment because, against orders, one woman induced one man to eat an apple? Special plead: you don't understand the subtle Doctrine of Free Will. Or: How can there be an equally godlike Father, Son, and Holy Ghost in the same Person? Special plead: You don't understand the Divine Mystery of the Trinity. Or: How could God permit the followers of Judaism, Christianity, and Islam—each in their own way enjoined to heroic measures of loving kindness and compassion—to have perpetrated so much cruelty for so long? Special plead: You don't understand Free Will again. And anyway, God moves in mysterious ways.*)

- **begging the question**, also called assuming the answer (e.g., *We must institute the death penalty to discourage violent crime.* But does the violent crime rate in fact fall when the death penalty is imposed? Or: *The stock market fell yesterday because of a technical adjustment and profit-taking by investors*—but is there any *independent* evidence for the causal role of "adjustment" and profit-taking; have we learned anything at all from this purported explanation?);

- **observational selection**, also called the enumeration of favorable circumstances, or as the philosopher Francis Bacon described it, counting the hits and forgetting the misses[3] (e.g., *A state boasts of the Presidents it has produced, but is silent on its serial killers*);

- **statistics of small numbers**—a close relative of observational selection (e.g., *"They say 1 out of every 5 people is Chinese. How is this possible? I know hundreds of people, and none of them is Chinese. Yours truly." Or: "I've thrown three sevens in a row. Tonight I can't lose."*);

- **misunderstanding of the nature of statistics** (e.g., *President Dwight Eisenhower expressing astonishment and alarm on discovering that fully half of all Americans have below average intelligence*);
- **inconsistency** (e.g., *Prudently plan for the worst of which a potential military adversary is capable, but thriftily ignore scientific projections on environmental dangers because they're not "proved." Or: Attribute the declining life expectancy in the former Soviet Union to the failures of communism many years ago, but never attribute the high infant mortality rate in the United States (now highest of the major industrial nations) to the failures of capitalism. Or: Consider it reasonable for the Universe to continue to exist forever into the future, but judge absurd the possibility that it has infinite duration into the past*);
- *non sequitur*—Latin for "It doesn't follow" (e.g., Our nation will prevail because God is great. But nearly every nation pretends this to be true; the German formulation was *"Gott mit uns"*). Often those falling into the *non sequitur* fallacy have simply failed to recognize alternative possibilities;
- *post hoc, ergo propter hoc*—Latin for "It happened after, so it was caused by" (e.g., *Jaime Cardinal Sin, Archbishop of Manila: "I know of . . . a 26-year-old who looks 60 because she takes [contraceptive] pills." Or: Before women got the vote, there were no nuclear weapons*);
- **meaningless question** (e.g., *What happens when an irresistible force meets an immovable object?* But if there is such a thing as an irresistible force there can be no immovable objects, and vice versa);
- **excluded middle, or false dichotomy**—considering only the two extremes in a continuum of intermediate possibilities (e.g., *"Sure, take his side; my husband's perfect; I'm always wrong." Or: "Either you love your country or you hate it." Or: "If you're not part of the solution, you're part of the problem"*);
- **short-term vs. long-term**—a subset of the excluded middle, but so important I've pulled it out for special attention (e.g., *We can't afford programs to feed malnourished children and educate pre-school kids. We need to urgently deal with crime on the streets. Or: Why explore space or pursue fundamental science when we have so huge a budget deficit?*);
- **slippery slope**, related to excluded middle (e.g., *If we allow abortion in the first weeks of pregnancy, it will be impossible to prevent the killing of a full-term infant. Or, conversely: If the state prohibits abortion even in the ninth month, it will soon be telling us what to do with our bodies around the time of conception*);
- **confusion of correlation and causation** (e.g., *A survey shows that more college graduates are homosexual than those with lesser education; therefore education makes people gay. Or: Andean earthquakes are correlated with closest approaches of the planet*

*Uranus; therefore—despite the absence of any such correlation for the nearer, more massive planet Jupiter—the latter causes the former[4]*);

- **straw man**—caricaturing a position to make it easier to attack (e.g., *Scientists suppose that living things simply fell together by chance*—a formulation that willfully ignores the central Darwinian insight, that Nature ratchets up by saving what works and discarding what doesn't. Or—this is also a short-term/long-term fallacy—*environmentalists care more for snail darters and spotted owls than they do for people*);

- **suppressed evidence,** or half-truths (e.g., *An amazingly accurate and widely quoted "prophecy" of the assassination attempt on President Reagan is shown on television;* but—an important detail—was it recorded before or after the event? Or: *These government abuses demand revolution, even if you can't make an omelette without breaking some eggs.* Yes, but is this likely to be a revolution in which far more people are killed than under the previous regime? What does the experience of other revolutions suggest? Are all revolutions against oppressive regimes desirable and in the interests of the people?);

- **weasel words** (e.g., The separation of powers of the U.S. Constitution specifies that the United States may not conduct a war without a declaration by Congress. On the other hand, Presidents are given control of foreign policy and the conduct of wars, which are potentially powerful tools for getting themselves re-elected. Presidents of either political party may therefore be tempted to arrange wars while waving the flag and calling the wars something else—"police actions," "armed incursions," "protective reaction strikes," "pacification," "safeguarding American interests," and a wide variety of "operations," such as "Operation Just Cause." Euphemisms for war are one of a broad class of reinventions of language for political purposes. Talleyrand said, "An important art of politicians is to find new names for institutions which under old names have become odious to the public").

Knowing the existence of such logical and rhetorical fallacies rounds out our toolkit. Like all tools, the baloney detection kit can be misused, applied out of context, or even employed as a rote alternative to thinking. But applied judiciously, it can make all the difference in the world—not least in evaluating our own arguments before we present them to others.


# NOTES


**Excerpted from Carl Sagan,** *The Demon-Haunted World: Science as a Candle in the Dark* **(New York: Random House, 1995), Ch. 12.**

1. This is a problem that affects jury trials. Retrospective studies show that some jurors make up their minds very early—perhaps during opening arguments—and then retain the evidence that seems to support their initial impressions and reject the contrary evidence. The method of alternative working hypotheses is not running in their heads.

2. A more cynical formulation by the Roman historian Polybius:

> Since the masses of the people are inconstant, full of unruly desires, passionate, and reckless of consequences, they must be filled with fears to keep them in order. The ancients did well, therefore, to invent gods, and the belief in punishment after death.

3. My favorite example is this story, told about the Italian physicist Enrico Fermi, newly arrived on American shores, enlisted in the Manhattan nuclear weapons Project, and brought face-to-face in the midst of World War II with U.S. flag officers:

So-and-so is a great general, he was told.

What is the definition of a great general? Fermi characteristically asked.

I guess it's a general who's won many consecutive battles.

How many?

After some back and forth, they settled on five.

What fraction of American generals are great?

After some more back and forth, they settled on a few percent.

But imagine, Fermi rejoined, that there is no such thing as a great general, that all armies are equally matched, and that winning a battle is purely a matter of chance. Then the chance of winning one battle is one out of two, or 1/2, two battles 1/4, three 1/8, four 1/16, and five consecutive battles 1/32—which is about 3 percent. You would expect a few percent of American generals to win five consecutive battles—purely by chance. Now, has any of them won ten consecutive battles . . . ?

4. Children who watch violent TV programs tend to be more violent when they grow up. But did the TV cause the violence, or do violent children preferentially enjoy watching violent programs? Very likely both are true. Commercial defenders of TV violence argue that anyone can distinguish between television and reality. But Saturday morning children's programs now average 25 acts of violence per hour. At the very least this desensitizes young children to aggression and random cruelty. And if impressionable adults can have false memories implanted in their brains, what are we implanting in our children when we expose them to some 100,000 acts of violence before they graduate from elementary school?

# 2

# WHY SMART PEOPLE BELIEVE WEIRD THINGS

## Michael Shermer

When men wish to construct or support a theory, how they torture facts into their service!

John Mackay,
*Extraordinary Popular Delusions*
*and the Madness of Crowds,* 1852

During the month of April 1998, when I was on a lecture tour for the first edition of my book *Why People Believe Weird Things*, the psychologist Robert Sternberg (best known for his pioneering work in multiple intelligences) attended my presentation at the Yale Law School. His response to the lecture was both enlightening and troubling. It is certainly entertaining to hear about *other* people's weird beliefs, Sternberg reflected, because we are confident that *we* would never be so foolish as to believe in such nonsense as alien abductions, ghosts, Bigfoot, ESP, and all manner of paranormal ephemera. But, he retorted, the interesting question is not why other people believe weird things, but why you and I believe weird things; and, as a subset of Us (versus Them), why *smart* people believe weird things. Sternberg then preceded to rattle off a number of beliefs held by his colleagues in psychology—by all accounts a reasonably smart cohort—that might reasonably be considered weird. And, he wondered with wry irony, which of his own beliefs . . . and mine . . . would one day be considered weird?

Unfortunately, there is no formal definition of a weird thing that most people can agree upon, because it depends so much on the particular claim being made in the context of the knowledge base that surrounds it and the individual or community proclaiming it. One person's weird belief might be another's normal theory, and a weird belief at one time might subsequently become normal. Stones falling from the sky were once the belief of a few daffy Englishmen; today we have an accepted theory of meteorites. In the jargon of science philosopher Thomas Kuhn, revolutionary ideas that are initially anathema to the accepted paradigm, in time may become normal science as the field undergoes a paradigm shift.[1]

Still, we can formulate a general outline of what might constitute a weird thing as we consider specific examples. For the most part, what I mean by a "weird thing" is: (1) a claim unaccepted by most experts in that particular field of study, (2) a claim that is either logically impossible or highly unlikely, and/or (3) a claim for which the evidence is largely anecdotal and uncorroborated. Most theologians, for example, recognize that God's existence cannot be proven in any scientific sense, and thus I consider William Dembski's Intelligent Design Theory, Hugh Ross's "Reasons to Believe," and Frank Tipler's Omega Point Theory—all of which purportedly use science to prove God—as not only unacceptable to most members of their knowledge community but as uncorroborated because such proof is logically impossible.[2]

"Smart people" suffers from a similar problem in operational definition, but at least here our task is aided by achievement criteria that most would agree, and the research shows, requires a minimum level of intelligence. Graduate degrees (especially the Ph.D.), university positions (especially at recognized and reputable institutions), peer-reviewed publications, and the like, allow us to concur that, while we might quibble over *how* smart some of these people are, the problem of smart people believing weird things is a genuine one that is quantifiable through measurable data.

## An Easy Answer to a Hard Question

It is a given assumption in the skeptical movement—elevated to a maxim really—that intelligence and education serve as an impenetrable prophylactic against the flim flam that we assume the unintelligent and uneducated masses swallow with credulity. Indeed, at the Skeptics Society we invest considerable resources in educational materials distributed to schools and the media under the assumption that this will make a difference in our struggle against pseudoscience and superstition. These efforts do make a difference, particularly for those who are aware of the phenomena we study but have not heard a scientific explanation for them. But are the cognitive elite protected against the nonsense that passes for sense in our culture? Is flapdoodle the fodder for only fools? The answer is no. The question is why?

For those of us in the business of debunking bunk and explaining the unexplained, this is what I call the Hard Question: *why do smart people believe weird things?* My Easy Answer will seem somewhat paradoxical at first:

*Smart people believe weird things because they are skilled at defending beliefs they arrived at for non-smart reasons.*

That is to say, most of us most of the time come to our beliefs for a variety of reasons having little to do with empirical evidence and logical reasoning (that, presumably, smart people are better at employing). Rather, such variables as genetic predispositions, parental predilections, sibling influences, peer pressures, educational experiences, and life impressions all shape the personality preferences and emotional inclinations that, in conjunction with numerous social and cultural influences, lead us to make certain belief choices. Rarely do any of us sit down before a table of facts, weigh them pro and con, and choose the most logical and rational belief, regardless of what we previously believed. Instead, the facts of the world come to us through the colored filters of the theories, hypotheses, hunches, biases, and prejudices we have accumulated through our lifetime. We then sort through the body of data and select those most confirming what we already believe, and ignore or rationalize away those that are disconfirming.

All of us do this, of course, but smart people are better at it through both talent and training. Some beliefs really are more logical, rational, and supported by the evidence than others, of course, but it is not my purpose here to judge the validity of beliefs; rather, I am interested in the question of how we came to them in the first place, and how we hold on to them in the face of either no evidence or contradictory evidence.

## The Psychology of Belief

There are a number of principles of the psychology of belief that go to the heart of fleshing out my Easy Answer to the Hard Question.

### 1. Intelligence and Belief.

Although there is some evidence that intelligent people are slightly less likely to believe in some superstitions and paranormal beliefs, overall conclusions are equivocal and limited. A study conducted in 1974 with Georgia high-school seniors, for example, found that those who scored higher on an I.Q. test were significantly less superstitious than students with lower I.Q. scores.[3] A 1980 study by psychologists James Alcock and L. P. Otis found that belief in various paranormal phenomena was correlated with lower critical thinking skills.[4] In 1989, W. S. Messer and R. A. Griggs found that belief in such psi phenomena as out-of-body experiences, ESP, and precognition was negatively correlated with classroom performance as measured by grades (as belief goes up, grades go down).[5]

But it should be noted that these three studies are using three different measures: I.Q., critical thinking skills, and educational performance. These may not always be indicative of someone being "smart." And what we mean by "weird things" here is not strictly limited to superstition and the paranormal. For example, cold fusion, creationism, and Holocaust revisionism could not reasonably be classified as superstitions or paranormal phenomena. In his review of the literature in one of the best books on this subject (*Believing in Magic*), psychologist Stuart Vyse concludes that while the relationship between intelligence and belief holds for some populations, it can be just the opposite in others. He notes that the New Age movement in particular "has led to the increased popularity of these ideas among groups previously thought to be immune to superstition: those with higher intelligence, higher socioeconomic status, and higher educational levels. As a result, the time-honored view of believers as less intelligent than nonbelievers may only hold for certain ideas or particular social groups."[6]

For the most part intelligence is orthogonal to and independent of belief. In geometry, orthogonal means "at right angles to something else"; in psychology orthogonal means "statistically independent of an experimental design: such that the variates under investigation can be treated as statistically independent," for example, "the concept that creativity and intelligence are relatively orthogonal (i.e., unrelated statistically) at high levels of intelligence."[7] Intuitively it seems as if the more intelligent people are the more creative they will be. In fact, in almost any profession significantly affected by intelligence (e.g., science, medicine, the creative arts), once you are at a certain level among the population of practitioners (and that level appears to be an I.Q. score of about 125), there is no difference in intelligence between the most successful and the average in that profession. At that point other variables take over, such as creativity, or achievement motivation and the drive to succeed, which are independent of intelligence.[8]

Cognitive psychologist Dean Keith Simonton's research on genius, creativity, and leadership, for example, has revealed that the raw intelligence of creative geniuses and leaders is not as important as their ability to generate a lot of ideas and select from them those that are most likely to succeed. Simonton argues that creative genius is best understood as a Darwinian process of variation and selection. Creative geniuses generate a massive variety of ideas from which they select only those most likely to survive and reproduce. As the two-time Nobel laureate and scientific genius Linus Pauling observed, one must "have lots of ideas and throw away the bad ones&. You aren't going to have good ideas unless you have lots of ideas and some sort of principle of selection." Like Forest Gump, genius is as genius does, says Simonton: "these are individuals credited with creative ideas or products that have left a large impression on a particular domain of intellectual or aesthetic activity. In other words, the creative genius attains eminence by leaving for posterity an impressive body of contributions that are both original and adaptive. In fact, empirical studies have repeatedly shown that the single most powerful predictor of eminence within any creative domain is the sheer number of influential products an individual has given the world." In science, for example, the number one predictor for receiving the Nobel Prize is the rate of journal citation—a measure, in part, of one's productivity. As well, Simonton notes, Shakespeare is a literary genius not just because he was good, but because "probably only the Bible is more likely to be found in English-speaking homes than is a volume containing the complete works of Shakespeare." In music, Simonton notes: "Mozart is considered a greater musical genius than Tartini in part because the former accounts for 30 times as much music in the classical repertoire as does the latter. Indeed, almost a fifth of all classical music performed in modern times was written by just three composers: Bach, Mozart, and Beethoven."[9] In other words, it is not so much that these creative geniuses were smart, but that they were productive and selective.

So intelligence is also orthogonal to the variables that go into shaping someone's beliefs. Another problem is that smart people might be smart in only one field. We say that their intelligence is domain specific. In the field of intelligence studies there is a long-standing debate about whether the brain is "domain general" or "domain specific." Evolutionary psychologists John Tooby, Leda Cosmides, and Steve Pinker, for example, reject the idea of a domain-general processor, focusing on brain modules that evolved to solve specific problems in our evolutionary history. On the other hand, many psychologists accept the notion of a global intelligence that could be considered domain general.[10] Archaeologist Steven Mithen goes so far as to say that it was a domain-general processor that made us human: "The critical step in the evolution of the modern mind was the switch from a mind designed like a Swiss army knife to one with cognitive fluidity, from a specialized to a generalized type of mentality. This enabled people to design complex tools, to create art and believe in religious ideologies. Moreover, the potential for other types of thought which are critical to the modern world can be laid at the door of cognitive fluidity."[11] It seems reasonable to argue that the brain consists of both domain specific and domain general modules. David Noelle, of the Center for the Neural Basis of Cognition at the Carnegie Mellon University, informs me:

> Modern neuroscience has made it clear that the adult brain *does* contain functionally distinct circuits. As our understanding of the brain advances, however, we find that these circuits rarely map directly onto complex domains of human experience, such as "religion" or "belief." Instead, we find circuits for more basic things, such as recognizing our location in space, predicting when something good is going to happen (e.g., when we will be rewarded), remembering events from our own lives, and keeping focused on our current goal. Complex aspects of behavior, like religious practices, arise from the interaction of these systems—*not* from any one module.[12]

What happens when smart people are smart in one field (domain specificity) but are not smart in an entirely different field, especially if the new field is a fertile breeding ground for weird beliefs? When Harvard marine biologist Barry Fell jumped fields into archaeology and wrote a best-selling book about all the people who discovered America before Columbus (*America B.C.*), he was woefully unprepared and obviously unaware that archaeologists had already considered his hypotheses of who first discovered America (Egyptians, Greeks, Romans, Phoenicians, etc.) but rejected them for lack of credible evidence. This is a splendid example of the social aspects of science, and why being smart in one field does not make one smart in another. Science is a social process, where one is trained in a certain paradigm and works with others in the field. A community of scientists read the same journals, go to the same conferences, review each other's papers and books, and generally exchange ideas about the facts, hypotheses, and theories in that field. Through vast experience they know, fairly quickly, which new ideas stand a chance of succeeding and which are obviously wrong. Newcomers from other fields typically dive in with both feet without the requisite training and experience, and proceed to generate new ideas that they think—because of their success in their own field—will be revolutionary. Instead, they are usually greeted with disdain (or, more typically, simply ignored) by the professionals in the field. This is not because (as they usually think) insiders don't like outsiders (or that all great revolutionaries are persecuted or ignored), but because in most cases those ideas were considered years or decades before and rejected for perfectly legitimate reasons.

### 2. Gender and Belief.

In many ways the orthogonal relationship of intelligence and beliefs is not unlike that of gender and beliefs. With the surge of popularity of psychic mediums like John Edward, James Van Praagh, and Sylvia Browne, it has become obvious to observers, particularly among journalists assigned to cover them, that at any given gathering (usually at large hotel conference rooms holding several hundred people, each of whom paid several hundred dollars to be there), that the vast majority (at least 75%) are women. Understandably, journalists inquire whether women, therefore, are more superstitious or less rational than men, who typically disdain such mediums and scoff at the notion of talking to the dead. Indeed, a number of studies have found that women hold more superstitious beliefs and accept more paranormal phenomena as real than men. In one study of 132 men and women in New York City, for example, scientists found that more women than men believed in knocking on wood or that walking under a ladder brought bad luck.[13] Another study showed that more college women than men professed belief in precognition.[14]

Although the general conclusion from such studies seems compelling, it is wrong. The problem here is with limited sampling. If you attend any meeting of creationists, Holocaust "revisionists," or UFOlogists, for instance, you will find almost no women at all (the few that I see at such conferences are the spouses of attending members and, for the most part, they look bored out of their skulls). For a variety of reasons related to the subject matter and style of reasoning, creationism, revisionism, and UFOlogy are guy beliefs. So, while gender is related to the *target* of one's beliefs, it appears to be unrelated to the process of believing. In fact, in the same study that found more women than men believed in precognition, it turns out that more men than women believe in Bigfoot and the Loch Ness Monster.[15] Seeing into the future is a woman's thing, tracking down chimerical monsters is a man's thing. There are no differences between men and women in the power of belief, only in what they choose to believe.

*3. Age and Belief.*

The relationship between age and belief is also mixed. Some studies show that older people are more skeptical than younger people, such as a 1990 Gallup poll indicating that people under thirty were more superstitious than older age groups.[16] Another study showed that younger police officers were more likely to believe in the full-moon effect (where allegedly crime rates are higher during the full moon) than older police officers. Other studies are less clear about the relationship. British folklorist Gillian Bennett discovered that older retired English women were more likely to believe in premonition than younger women.[17] Psychologist Seymour Epstein surveyed three different age groups (9–12, 18–22, 27–65) and discovered that the percentage of belief in each age division depended on the specific phenomena under question. For telepathy and precognition there were no age group differences. For good luck charms more older adults said they had one than did college students or children. The belief that wishing something to happen will make it so decreased steadily with age.[18] Finally, Frank Sulloway and I found that religiosity and belief in God steadily decreased with age, until about age seventy-five, when it went back up.[19]

These mixed results are due to what is known as person-by-situation effects, where a simple linear causal relationship between two variables rarely exists. Instead, to the question "does X cause Y?" the answer is often "it depends." Bennett, for example, concluded that the older women in her study had lost power, status, and especially, had lost loved ones. Belief in the supernatural helped them deal with these losses. Sulloway and I concluded in our study that age and religiosity varies according to one's situation in relation to both early powerful influences and the later perceived impending end of life.

*4. Education and Belief.*

Studies on the relationship between education and belief are, like intelligence, gender, and age, mixed. Psychologist Chris Brand, for example, discovered a powerful inverse correlation of −.50 between I.Q. and authoritarianism (as I.Q. increases authoritarianism decreases). Brand concluded that authoritarians are characterized not by affection for authority, but by "some simple-minded way in which the world has been divided up for them." In this case, authoritarianism was being expressed through prejudice by dividing the world up by race, gender, and age. Brand attributes the correlation to "crystallized intelligence," a relatively flexible form of intelligence shaped by education and life experience. But Brand is quick to point out that only when this type of intelligence is modified by a liberal education does one see a sharp decrease in authoritarianism. In other words, it is not so much that *smart* people are less prejudiced and authoritarian, but that *educated* people are less so.[20]

Psychologists S. H. and L. H. Blum found a negative correlation between education and superstition (as education increased, superstitious beliefs decreased).[21] Laura Otis and James Alcock showed that college professors are more skeptical than either college students or the general public (with the latter two groups showing no difference in belief), but that within college professors there was variation in the types of beliefs held, with English professors more likely to believe in ghosts, ESP, and fortune telling.[22] Another study found, not surprisingly, that natural and social scientists were more skeptical than their colleagues in the arts and humanities; most appropriately, in this context, psychologists were the most skeptical of all (perhaps because they best understand the psychology of belief and how easy it is to be fooled).[23]

Finally, Richard Walker, Steven Hoekstra, and Rodney Vogl discovered that there was no relationship between science education and belief in the paranormal among three groups of science students at three different colleges. That is, "having a strong scientific knowledge base is not enough to insulate a person against irrational beliefs. Students who scored well on these tests were no more or less skeptical of pseudoscientific claims than students who scored very poorly. Apparently, the students were not able to apply their scientific knowledge to evaluate these pseudoscientific claims. We suggest that this inability stems in part from the way that science is traditionally presented to students: Students are taught *what* to think but not *how* to think."[24]

Whether teaching students how to think will attenuate belief in the paranormal remains to be seen. Supposedly this is what the critical thinking movement has been emphasizing for three decades now, yet polls show that paranormal beliefs continue to rise. A June 8, 2001 Gallup Poll, for example, reported a significant increase in belief in a number of paranormal phenomena since 1990, including haunted houses, ghosts, witches, communicating with the dead, psychic or spiritual healing, that extraterrestrial beings have visited Earth, and clairvoyance. In support of my claim that the effects of gender, age, and education show content dependent effects, the Gallup poll found:

*Gender:* Women are slightly more likely than men to believe in ghosts and that people can communicate with the dead. Men, on the other hand, are more likely than women to believe in only one of the dimensions tested: that extraterrestrials have visited Earth at some point in the past.

*Age:* Younger Americans—those 18–29—are much more likely than those who are older to believe in haunted houses, in witches, in ghosts, that extraterrestrials have visited Earth, and in clairvoyance. There is little significant difference in belief in the other items by age group. Those 30 and older are somewhat more likely to believe in possession by the Devil than are the younger group.

*Education:* Americans with the highest levels of education are more likely than others to believe in the power of the mind to heal the body. On the other hand, belief in three of the phenomena tested goes up as the educational level of the respondent goes down: possession by the Devil, astrology and haunted houses.

Gallup reported an even more striking poll result on March 5, 2001, about the surprising lack of belief in and understanding of the theory of evolution. Specifically, of those Americans polled:

- 45% agreed with the statement: "God created human beings pretty much in their present form at one time within the last 10,000 years or so."
- 37% agreed with the statement: "Human beings have developed over millions of years from less advanced forms of life, but God guided this process."
- 12% agreed with the statement: "Human beings have developed over millions of years from less advanced forms of life, but God had no part in this process."

Despite enormous funds and efforts allocated toward the teaching of evolution in public schools, and the proliferation of documentaries, books, and magazines presenting the theory on all levels, Americans have not noticeably changed their opinion on this question since Gallup started asking it in 1982. Gallup did find that individuals with more education and people with higher incomes are more likely to think that evidence supports the theory of evolution, and that younger people are also more likely than older people to think that evidence supports Darwin's theory (again confounding the age variable). Nevertheless, only 34% of Americans consider themselves to be "very informed" about the theory of evolution, while a slightly greater percentage—40%—consider themselves to be "very informed" about the theory of creation. Younger people, people with more education, and people with higher incomes are more likely to say they are very informed about both theories.[26]

## 5. Personality and Belief.

Clearly human thought and behavior are complex and thus studies such as those reported above rarely show simple and consistent findings. Studies on the causes and effects of mystical experiences, for example, show mixed findings. The religious scholar Andrew Greeley and others have found a slight but significant tendency for mystical experiences to increase with age, education, and income, but there were no gender differences.[27] J. S. Levin, by contrast, in analyzing the 1988 General Social Survey data, found no significant age trends in mystical experiences.[28]

The psychologist David Wulff, in a general survey of the literature on the psychology of mystical experiences (a subset of weird things), concluded that there were some consistent personality differences:

> Persons who tend to score high on mysticism scales tend also to score high on such variables as complexity, openness to new experience, breadth of interests, innovation, tolerance of ambiguity, and creative personality. Furthermore, they are likely to score high on measures of ability to be hypnotized, absorption, and fantasy proneness, suggesting a capacity to suspend the judging process that distinguishes imaginings and real events and to commit their mental resources to representing the imagined object as vividly as possible.

Individuals high on hypnotic susceptibility are also more likely to report having undergone religious conversion, which for them is primarily an experiential rather than a cognitive phenomenon—that is, one marked by notable alterations in perceptual, affective, and ideomotor response patterns.[29]

### 6. Locus of Control and Belief.

One of the most interesting areas of research on the psychology of belief is in the area of what psychologists call *locus of control*. People who measure high on *external locus of control* tend to believe that circumstances are beyond their control and that things just happen to them. People who measure high on *internal locus of control* tend to believe they are in control of their circumstances and that they make things happen.[30] External locus of control leads to greater anxiety about the world, whereas internal locus of control leads one to be more confident in one's judgment, skeptical of authority, and less compliant and conforming to external influences. In relation to beliefs, studies show that skeptics are high in internal locus of control whereas believers are high in external locus of control.[31] A 1983 study by Jerome Tobacyk and Gary Milford of introductory psychology students at Louisiana Tech University, for example, found that those who scored high in external locus of control tended to believe in ESP, witchcraft, spiritualism, reincarnation, precognition, and were more superstitious than those students who scored high in internal locus of control.[32]

James McGarry and Benjamin Newberry in a 1977 study of strong believers in and practitioners of ESP and psychic power, however, found an interesting twist to this effect. Surprisingly, this group scored high in internal locus of control. The authors offered this explanation: "These beliefs [in ESP] may render such a person's problems less difficult and more solvable, lessen the probability of unpredictable occurrences, and offer hope that political and governmental decisions can be influenced."[33] In other words, a deep commitment to a belief in ESP, which usually entails believing that the individual themselves has ESP, changes the focus from external to internal locus of control.

The environment also mitigates the effect of locus of control on belief, where there is a relationship between the uncertainty of an environment and the level of superstitious belief (as uncertainty goes up so too do superstitions). The anthropologist Bronislaw Malinowski, for example, discovered that among the Trobriand Islanders (off the coast of New Guinea), the further out to sea they went to fish the more they developed superstitious rituals. In the calm waters of the inner lagoon, there were very few rituals. By the time they reached the dangerous waters of deep-sea fishing, the Trobrianders were also deep into magic. Malinowski concluded that magical thinking derived from environmental conditions, not inherent stupidities: "We find magic wherever the elements of chance and accident, and the emotional play between hope and fear have a wide and extensive range. We do not find magic wherever the pursuit is certain, reliable, and well under the control of rational methods and technological processes. Further, we find magic where the element of danger is conspicuous."[34] Think of the superstitions of baseball players. Hitting a baseball is exceedingly difficult, with the best succeeding barely more than three out of every ten times at bat. And hitters are known for their extensive reliance on rituals and superstitions that they believe will bring them good luck. These same superstitious players, however, drop the superstitions when they take the field, since most of them succeed in fielding the ball over 90 percent of the time. Thus, as with the other variables that go into shaping belief that are themselves orthogonal to intelligence, the context of the person and the belief system are important.

### 7. Influence and Belief.

Scholars who study cults (or, as many prefer to call them by the less pejorative term, "New Religious Movements") explain that there is no simple answer to the question "who joins cults?" The only consistent variable seems to be age—young people are more likely to join cults than older people—but beyond that, variables such as family background, intelligence, and gender are orthogonal to belief in and commitment to cults. Research shows that two-thirds of cult members come from normal functioning families and showed no psychological abnormalities whatsoever when they joined the cult.[35] Smart people and non-smart people both readily join cults, and while women are more likely to join such groups as J. Z. Knight's "Ramtha"-based cult (she allegedly channels a 35,000-year old guru named "Ramtha" who doles out life wisdom and advice, in English with an Indian accent no less!), men are more likely to join militias and other anti-government groups.

Again, although intelligence may be related to how well one is able to justify one's membership in a group, and while gender may be related to which group is chosen for membership, intelligence and gender are unrelated to the general process of joining, the desire for membership in a cult, and belief in the cult's tenets. Psychiatrist Marc Galanter, in fact, suggests that joining such groups is an integral part of the human condition to which we are all subject due to our common evolutionary heritage.[36] Bonding together in closely-knit groups was a common practice in our evolutionary history, because it reduced risk and increased survival by being with others of our perceived kind. But if the process of joining is common among most humans, why do some people join while others do not?

The answer is in the persuasive power of the principles of influence and the choice of what type of group to join. Cult experts and activists Steve Hassan and Margaret Singer outline a number of psychological influences that shape the thoughts and behaviors that lead people to join the more dangerous groups (and that are quite independent of intelligence): cognitive dissonance, obedience to authority, group compliance and conformity, and especially the manipulation of rewards, punishments, and experiences with the purpose of controlling behavior, information, thought, and emotion (what Hassan calls the "BITE model").[37]

Social psychologist Robert Cialdini demonstrates in his enormously persuasive book on influence, that all of us are influenced by a host of social and psychological variables, including physical attractiveness, similarity, repeated contact or exposure, familiarity, diffusion of responsibility, reciprocity, and many others.[38]

## Smart Biases in Defending Weird Beliefs

In 1620 English philosopher and scientist Francis Bacon offered his own Easy Answer to the Hard Question:

> The human understanding when it has once adopted an opinion (either as being the received opinion or as being agreeable to itself) draws all things else to support and agree with it. And though there be a greater number and weight of instances to be found on the other side, yet these it either neglects and despises, or else by some distinction sets aside and rejects; in order that by this great and pernicious predetermination the authority of its former conclusions may remain inviolate. . . . And such is the way of all superstitions, whether in astrology, dreams, omens, divine judgments, or the like; wherein men, having a delight in such vanities, mark the events where they are fulfilled, but where they fail, although this happened much oftener, neglect and pass them by.[39]

Why do smart people believe weird things? Because, to restate my thesis in light of Bacon's insight, *smart people believe weird things because they are skilled at defending beliefs they arrived at for non-smart reasons.*

As we have already seen, there is a wealth of scientific evidence in support of this thesis, but none more so than an extremely powerful cognitive bias that makes it difficult for any of us to objectively evaluate a claim.

*The Confirmation Bias.* At the core of the Easy Answer to the Hard Question is the confirmation bias, or the tendency to seek or interpret evidence favorable to already existing beliefs, and to ignore or reinterpret evidence unfavorable to already existing beliefs. Psychologist Raymond Nickerson, in a comprehensive review of the literature on this bias, concluded: "If one were to attempt to identify a single problematic aspect of human reasoning that deserves attention above all others, the confirmation bias would have to be among the candidates for consideration . . . it appears to be sufficiently strong and pervasive that one is led to wonder whether the bias, by itself, might account for a significant fraction of the disputes, altercations, and misunderstandings that occur among individuals, groups, and nations."[40]

Although lawyers deliberately employ a type of confirmation bias in the confrontational style of reasoning used in the courtroom by purposefully selecting evidence that best suits their client and ignoring contradictory evidence (where winning the case trumps the truth or falsity of the claim), psychologists believe that, in fact, we all do this, usually unconsciously. In a 1989 study psychologists Bonnie Sherman and Ziva Kunda presented students with evidence that contradicted a belief they held deeply, as well as with evidence that supported those same beliefs, it was observd that the students tended to attenuate the validity of the first set of evidence and accentuate the value of the second.[41] In a 1989 study with both children and young adults who were exposed to evidence inconsistent with a theory they preferred, Deanna Kuhn found that they "either failed to acknowledge discrepant evidence or attended to it in a selective, distorting manner. Identical evidence was interpreted one way in relation to a favored theory and another way in relation to a theory that was not favored."[42] After the experiment subjects could not even remember what the contradictory evidence was. In a subsequent study in 1994 Kuhn exposed subjects to an audio recording of an actual murder trial and discovered that instead of evaluating the evidence objectively, most subjects first composed a story of what happened, and then sorted through the evidence to see what best fit that story.[43] Interestingly, those subjects most focused on finding evidence for a single view of what happened (as opposed to those subjects willing to at least consider an alternative scenario), were the most confident in their decision.

Even in judging something as subjective as personality, psychologists have found that we see what we are looking for in a person. In a series of studies, subjects were asked to assess the personality of someone they were about to meet. Some were given a profile of an introvert (shy, timid, quiet), others given a profile of an extrovert (sociable, talkative, outgoing). When asked to make a personality assessment, those told that the person would be an extrovert asked questions that would lead to that conclusion; the introvert group did the same. They both found in the person the personality they were seeking to find.[44] Of course, the confirmation bias works both ways in this experiment. It turns out that the subjects whose personalities were being evaluated tended to give answers that would confirm whatever hypothesis the interrogator was holding.

The confirmation bias is not only pervasive, but its effects can be powerfully influential on people's lives. In a 1983 study John Darley and Paul Gross showed subjects a video of a child taking a test. One group was told that the child was from a high socioeconomic class while the other group was told that the child was from a low socioeconomic class. The subjects were then asked to evaluate the academic abilities of the child based on the results of the test. Not surprisingly, the high socioeconomic group rated the child's abilities as above grade level, while the low socioeconomic group rated the child's abilities as below grade level. In other words, the same data were seen by one group of evaluators differently from the other group, depending on what their expectations were. The data then confirmed those expectations.[45]

The confirmation bias can also influence one's emotional states and prejudices. Hypochondriacs interpret every little ache and pain as indications of the next great health calamity, whereas normal people simply ignore such random bodily signals.[46] Paranoia is another form of confirmation bias, where if you strongly believe that "they" are out to get you then you will interpret the wide diversity of anomalies and coincidences in life to be evidence in support of that paranoid hypothesis. Likewise, prejudice depends on a type of confirmation bias, where the prejudged expectations of a group's characteristics leads one to evaluate an individual who is a member of that group in terms of those expectations.[47] When depressed people tend to focus on those events and information that further reinforce the depression, and to suppress evidence that things are, in fact, getting better.[48] As Nickerson noted in summary: "the presumption of a relationship predisposes one to find evidence of that relationship, even when there is none to be found or, if there is evidence to be found, to overweight it and arrive at a conclusion that goes beyond what the evidence justifies."[49]

Even scientists are subject to the confirmation bias. Often in search of a particular phenomenon, scientists interpreting data may see (or select) those data most in support of the hypothesis under question and ignore (or toss out) those data not in support of the hypothesis. Historians of science have determined, for example, that the confirmation bias was hard at work in one of the most famous experiments in the history of science. In 1919 the British astronomer Arthur Stanley Eddington tested Einstein's prediction for how much the sun would deflect light coming from a background star during an eclipse (the only time you can see stars behind the sun). It turns out that Eddington's measurement error was as great as the effect he was measuring. As Stephen Hawking described it, "The British team's measurement had been sheer luck, or a case of knowing the result they wanted to get, not an uncommon occurrence in science."[50] In going through Eddington's original data, historians S. Collins and J. Pinch found that "Eddington could only claim to have confirmed Einstein because he used Einstein's derivations in deciding what his observations really were, while Einstein's derivations only became accepted because Eddington's observation seemed to confirm them. Observation and prediction were linked in a circle of mutual confirmation rather than being independent of each other as we would expect according to the conventional idea of an experiment test."[51] In other words, Eddington found what he was looking for. Of course, science contains a special self-correcting mechanism to get around the confirmation bias, and that is that other people will check your results or rerun the experiment. If your results were entirely the product of the confirmation bias, someone will sooner or later catch you on it. That is what sets science apart from all other ways of knowing.

Finally, and most important for our purposes here, the confirmation bias operates to confirm and justify weird beliefs. Psychics, fortunetellers, palm readers, and astrologers, for example, all depend on the power of the confirmation bias by telling their clients (some would call them "marks") what to expect in their future. By offering them one-sided events (instead of two-sided events in which more than one outcome is possible), the occurrence of the event is noticed while the nonoccurrence of the event is not. Consider numerology. The search for meaningful relationships in various measurements and numbers available in almost any structure in the world (including the world itself, as well as the cosmos), has led numerous observers to find deep meaning in the relationship between these numbers. The process is simple. You can start off with the number you seek and try to find some relationship that ends in that number, or one close to it. Or, more commonly, you crunch through the numbers and see what pops out of the data that looks familiar. In the Great Pyramid, for example, the ratio of the pyramid's base to the width of a casing stone is 365, the number of days in the year. Such number crunching with the confirmation bias in place has led people to "discover" in the pyramid the earth's mean density, the period of precession of the earth's axis, and the mean temperature of the earth's surface. As Martin Gardner wryly noted, this is a classic example of "the ease with which an intelligent man, passionately convinced of a theory, can manipulate his subject matter in such a way as to make it conform to precisely held opinions."[52] And the more intelligent the better.

So, in sum, being either high or low in intelligence is orthogonal to and independent of the normalness or weirdness of beliefs one holds. But these variables are not without some interaction effects. High intelligence, as noted in my Easy Answer, makes one *skilled at defending beliefs arrived at for non-smart reasons.*

## NOTES

**Originally published in** *Skeptic* **10, No. 2 (September 15, 2003): 62–73.**

1. Kuhn, T. 1962. *The Structure of Scientific Revolutions.* Chicago: University of Chicago Press. See also Kuhn, T. 1977. *The Essential Tension: Selected Studies in Scientific Tradition and Change.* Chicago: University of Chicago Press.

2. Dembski, W. 1998. *The Design Inference: Eliminating Chance Through Small Probabilities.* Cambridge: Cambridge University Press; Ross, H. 1993. *The Creator and the Cosmos: How the Greatest Scientific Discoveries of the Century Reveal God.* Colorado Springs, CO: Navpress; 1994. Creation and Time: A Biblical and Scientific Perspective on the Creation-Date Controversy. Colorado Springs, CO: Navpress; 1996. *Beyond the Cosmos: What Recent Discoveries in Astronomy and Physics Reveal About the Nature of God.* Colorado Springs, CO: Navpress; Tipler, F. 1994. *The Physics of Immortality: Modern Cosmology, God and the Resurrection of the Dead.* New York: Doubleday.

3. Killeen, P., R. W. Wildman, and R. W. Wildman II. 1974. "Superstitiousness and Intelligence." *Psychological Reports.* 34, 1158.

4. Alcock, J. E. and Otis, L. P. 1980. "Critical Thinking and Belief in the Paranormal." *Psychological Reports.* 46, 479–482.

5. Messer, W. S. and R. A. Griggs. 1989. "Student Belief and Involvement in the Paranormal and Performance in Introductory Psychology." *Teaching of Psychology.* 16, 187–191.

6. Vyse, S. A. 1997. *Believing in Magic: the Psychology of Superstition.* New York: Oxford University Press, 47.

7. *Oxford English Dictionary.* The discussion of the orthogonal relationship between creativity and intelligence can be found in 1973 *Journal of Genetic Psychology*, 72: 45

8. Hudson, L. 1966. *Contrary Imaginations: A Psychological Study of the English Schoolboy.* London: Methuen; Getzels, J. W. and P. W. Jackson. 1962. *Creativity and Intelligence: Explorations with Gifted Students.* New York: John Wiley.

9. Simonton, D. K. 1999. *Origins of Genius: Darwinian Perspectives on Creativity.* Oxford: Oxford University Press. See also Sulloway, F. 1996. *Born to Rebel: Birth Order, Family Dynamics, and Creative Lives.* New York: Pantheon.

10. Barkow, J. H., L. Cosmides, J. Tooby. 1992. *The Adapted Mind.* Oxford University Press; Pinker, S. 1997. *How the Mind Works.* New York: W. W. Norton.

11. Mithen, S. 1996. T*he Prehistory of the Mind: The Cognitive Origins of Art, Religion, and Science.* London: Thames and Hudson, 163; see also Jensen, A. R. 1998. *The g Factor: The Science of Mental Ability.* Westport, CT: Praeger. For multiple intelligences see: Gardner, H. 1983. *Frames of Mind: The Theory of Multiple Intelligences.* New York: Basic Books; and Sternberg, R. J. 1996. *Successful Intelligence: How Practical and Creative Intelligence Determine Succcess in Life.* New York: Simon and Schuster.

12. Personal correspondence, 3/19/98; for a critique of brain modularity see Karmiloff-Smith, A. 1995. *Beyond Modularity: A Developmental Perspective on Cognitive Science.* London: Bradford.

13. Blum, S. H. and L. H. Blum. 1974. "Do's and Don'ts: An Informal Study of Some Prevailing Superstitions." *Psychological Reports* 35, 567–571.

14. Tobacyk, J. and G. Milford. 1983. "Belief in Paranormal Phenomena: Assessment Instrument Development and Implications for Personality Functioning." *Journal of Personality and Social Psychology* 44, 1029–1037.

15. Ibid.

16. Bennett, G. 1987. *Traditions of Belief: Women, Folklore, and the Supernatural Today.* London: Penguin Books.

17. Epstein, S. 1993. "Implications of Cognitive-Experiential Self-Theory for Personality and Developmental Psychology." In D. C. Funder, et al. (Eds.). *Studying Lives Through Time: Personality and Developmental Psychology*, 399–438. Washington, D.C.: American Psychological Association.

18. Vyse, op. cit.

19. Shermer, M. and F. Sulloway. 2003. "Belief in God: An Empirical Study." Forthcoming.

20. Brand, C. 1981. "Personality and Political Attitudes." In R. Lynn (ed.), *Dimensions of Personality; Papers in Honour of H. J. Eysenck.* Oxford: Pergamon Press., 7–38, 28.

21. Blum, S. H. and L. H. Blum. 1974. "Do's and Don'ts: An Informal Study of Some Prevailing Superstitions." *Psychological Reports* 35, 567–571.

22. Otis, L. P. and J. E. Alcock. 1982. "Factors Affecting Extraordinary Belief." *The Journal of Social Psychology.* 118, 77–85.

23. Pasachoff, J. M., R. J. Cohen, and N. W. Pasachoff. 1971. "Belief in the Supernatural Among Harvard and West African University Students." *Nature*, 232, 278–279.

24. Walker, W. R., S. J. Hoekstra, and R. J. Vogl. 2001. "Science Education is No Guarantee for Skepticism." *Skeptic*, Vol. 9, No. 3.

25. The Gallup Poll report is at http://www.gallup.com/poll/releases/pr010608.asp.

26. Ibid.

27. Greeley, A. M. 1975. *The Sociology of the Paranormal: A Reconnaissance.* Beverly Hills, CA: Sage; Hay, D. and A. Morisy. 1978. "Reports of Ecstatic, Paranormal, or Religious Experience in Great Britain and the United States—A Comparison of Trends." *Journal for the Scientific Study of Religion*, 17, 255–268.

28. Levin, J. S. 1993. "Age Differences in Mystical Experience." *The Gerontologist*, 33, 507–513.

29. David M. Wulff, 2000. "Mystical Experience." In *Varieties of Anomalous Experience: Examining the Scientific Evidence.* (E. Cardena, S. J. Lynn, and S. Krippner, eds.). Washington, D.C.: American Psychological Association, 408.

30. Rotter, J. B. 1966. "Generalized Expectancies for Internal versus External Control of Reinforcement." *Psychological Monographs*, 80, 609: 1–28.

31. Marshall, et al., 1994. "The Five-Factor Model of Personality as a Framework for Personality-health Research." *Journal of Personality and Social Psychology*, 67, 278–286.

32. Tobacyk, J. and G. Milford. 1983. "Belief in Paranormal Phenomena: Assessment Instrument Development and Implications for Personality Functioning." *Journal of Personality and Social Psychology* 44, 1029–1037.

33. McGarry, J. JU. and B. H. Newberry. 1981. "Beliefs in Paranormal Phenomena and Locus of Control: A Field Study." *Journal of Personality and Social Psychology.* 41, 725–736.

34. Malinowski, B. 1954. *Magic, Science, and Religion.* New York: Doubleday, 139–140.

35. Singer, M. 1995. *Cults in Our Midst: The Hidden Menace in Our Everyday Lives.* San Francisco: Jossey-Bass Publishers.

36. Galanter, M. 1999. *Cults: Faith, Healing, and Coercion.* 2nd Edition. New York: Oxford University Press; Hassan, S. 1990. *Combatting Cult Mind Control.* Rochester, VT: Park Street Press.

37. Hassan, S. 2000. *Releasing the Bonds: Empowering People to Think for Themselves.* Somerville, MA: Freedom of Mind Press.

38. Cialdini, R. 1984. *Influence: The New Psychology of Modern Persuasion.* New York: William Morrow.

39. Bacon, F. 1620 (1939). *Novum Organum.* In Burtt, E. A. (ed.). *The English Philosophers from Bacon to Mill.* New York: Random House.

40. Nickerson, R. S. 1998. "Confirmation Bias: A Ubiquitous Phenomenon in Many Guises." *Review of General Psychology*, Vol. 2, No. 2: 175–220, 175.

41. Sherman, B. and Z. Kunda. 1989. "Motivated Evaluation of Scientific Evidence." Paper presented at the annual meeting of the American Psychological Society, Arlington, VA.

42. Kuhn, D. 1989. "Children and Adults as Intuitive Scientists." *Psychological Review*, 96, 674–689.

43. Kuhn, D., M. Weinstock, and R. Flaton. 1994. "How Well Do Jurors Reason? Competence Dimensions of Individual Variation in a Juror Reasoning Task." *Psychological Science*, 5, 289–296.

44. Snyder, M. 1981. "Seek and Ye Shall Find: Testing Hypotheses About Other People." In E. T. Higgins, C. P. Heiman, and M. P. Zanna (eds.), *Social Cognition: The Ontario Symposium on Personality and Social Psychology* (277–303). Hillsdale, NJ: Erlbaum.

45. Darley, J. M. and P. H. Gross. 1983. "A Hypothesis-Confirming Bias in Labelling Efects." *Journal of Personality and Social Psychology*, 44, 20–33.

46. Pennebaker, J. W. and J. A. Skelton. 1978. "Psychological Parameters of Physical symptoms." *Personality and Social Psychology Bulletin*, 4, 524–530.

47. Hamilton, D. L., P. M. Dugan, and T. K. Trolier. 1985. "The Formation of Stereotypic Beliefs: Further Evidence for Distinctiveness-Based Illusory Correlations." *Journal of Personality and Social Psychology*, 48, 5–17.

48. Beck, A. T. 1976. *Cognitive Therapy and the Emotional Disorders*. New York: International Universities Press.

49. Nickerson, op cit., 182–183.

50. Hawking, S. 1988. *A Brief History of Time*. New York: Bantam Books.

51. Collins, S., and J. Pinch. 1993. *The Golem: What Everyone Should Know About Science*. New York: Cambridge University Press.

52. Gardner, M. 1957. *Fads and Fallacies in the Name of Science*. New York: Dover.

# 3

## COLD READING:
## HOW TO CONVINCE STRANGERS THAT
## YOU KNOW ALL ABOUT THEM

### *Ray Hyman*

Over twenty years ago I taught a course at Harvard University called "Applications of Social Psychology." The sort of applications that I covered were the various ways in which people were manipulated. I invited various manipulators to demonstrate their techniques—pitchmen, encyclopedia salesmen, hypnotists, advertising experts, evangelists, confidence men, and a variety of individuals who dealt with personal problems. The techniques which we discussed, especially those concerned with helping people with their personal problems, seem to involve the client's tendency to find more meaning in any situation than is actually there. Students readily accepted this explanation when it was pointed out to them. But I did not feel that they fully realized just how pervasive and powerful this human tendency to make sense out of nonsense really is.

Consequently, in 1955 I wrote a paper entitled "The Psychological Reading: An Infallible Technique For Winning Admiration and Popularity." Over the years I have distributed copies of this paper to my students. The paper begins as follows:

> So you want to be admired! You want people to seek your company, to talk about you, to praise your talents. This manuscript tells you how to satisfy that want. Herein you will find a sure-fire gimmick for the achievement of fame and popularity. Just follow the advice that I give you, and, even if you are the most incompetent social bungler, you cannot fail to become the life of the party. What is the secret that underlies this infallible system! The secret, my friend, is a simple and obvious one. It has been tried and proven by practitioners since the beginnings of mankind. Here is the gist of the secret: To be popular with your fellow man, tell him what he wants to hear. He wants to hear about himself. So tell him about him self. But [do not tell him]

what you know to he true about him. Oh, no! Never tell him the
truth. Rather, tell him what he would like to hear about himself. And
there you have it. Simple and obvious, yet so powerful. This
manuscript details the way in which you can exploit this golden rule
by assuming the role of a character reader.

I will include essentially the same recipe for character reading in this paper
that I give to my students. In addition I will bring the material up to date,
describe some relevant research, and indicate some theoretical reasons why the
technique "works." My purpose is not to enable you to enhance your personal
magnetism, nor is it to increase the number of character readers. I give you these
rules for reading character because I want you to experience how the method
works. I want you to see what a powerful technique the psychological reading is,
how convincing it is to the psychologist and layman alike.

When you see how easy it is to convince a person that you can read his
character on sight, you will better appreciate why fortune tellers and
psychologists are frequently lulled into placing credence in techniques which
have not been validated by acceptable scientific methods. The recent
controversy in *The Humanist* magazine and *The Zetetic* over the scientific status
of astrology probably is irrelevant to the reasons that individuals believe in
astrology. Almost without exception the defenders of astrology with whom I
have contact do not refer to the evidence relating to the underlying theory. They
are convinced of astrology's value because it "works." By this they mean that it
supplies them with feedback that "feels right"—that convinces them that the
horoscope provides a basis for understanding themselves and ordering their
lives. It has personal meaning for them.

Some philosophers distinguish between "persuasion" and "conviction." The
distinction is subtle. But for our purposes we can think of subjective experiences
that persuade us that something is so and of logical and scientific procedures
that convince, or ought to convince, us that something is or is not so. Quite
frequently a scientist commits time and resources toward generating scientific
evidence for a proposition because he has already been persuaded, on
nonscientific grounds, that the proposition is true. Such intuitive persuasion
plays an important motivational role in science as well as in the arts.
Pathological science and false beliefs come about when such intuitive
persuasion overrides or colors the evidence from objective procedures for
establishing conviction.

The field of personality assessment has always been plagued by this
confusion between persuasion and conviction. In contrast to intelligence and
aptitude tests the scientific validation of personality tests, even under ideal
conditions, rarely results in unequivocal or satisfactory results. In fact some of
the most widely used personality inventories have repeatedly failed to pass
validity checks. One of the reasons for this messy state of affairs is the lack of
reliable and objective criteria against which to check the results of an
assessment.

But the lack of adequate validation has not prevented the use of, and reliance on, such instruments. Assessment psychologists have always placed more reliance on their instruments than is warranted by the scientific evidence. Both psychologist and client are invariably persuaded by the results that the assessment "works."

This state of affairs, of course, is even more true when we consider divination systems beyond those of the academic and professional psychologist. Every system be it based on the position of the stars, the pattern of lines in the hand, the shape of the face or skull, the fall of the cards or the dice, the accidents of nature, or the intuitions of a "psychic"—claims its quota of satisfied customers. The client invariably feels satisfied with the results. He is convinced that the reader and the system have penetrated to the core of his "true" self. Such satisfaction on the part of the client also feeds back upon the reader. Even if the reader began his career with little belief in his method, the inevitable reinforcement of persuaded clients increases his confidence in himself and his system. In this way a "vicious circle" is established. The reader and his clients become more and more persuaded that they have hold of a direct pipeline to the "truth."

The state of affairs in which the evaluation of an assessment instrument depends upon the satisfaction of the client is known as "personal validation." Personal validation is, for all practical purposes, the major reason for the persistence of divinatory and assessment procedures. If the client is not persuaded, then the system will not survive. Personal validation, of course, is the basis for the acceptance of more than just assessment instruments. The widespread acceptance of myths about Bigfoot, the Bermuda Triangle, ancient astronauts, ghosts, the validity of meditation and consciousness-raising schemes, and a host of other beliefs are based on persuasion through personal validation rather than scientific conviction.

## Cold Reading

"Cold reading" is a procedure by which a "reader" is able to persuade a client, whom he has never before met, that he knows all about the client's personality and problems. At one extreme this can be accomplished by delivering a stock spiel, or "psychological reading," that consists of highly general statements that can fit any individual. A reader who relies on psychological readings will usually have memorized a set of stock spiels. He then can select a reading to deliver which is relatively more appropriate in the general category that the client fits- a young unmarried girl, a senior citizen, and so on. Such an attempt to fit the reading to the client makes the psychological reading a closer approximation to the true cold reading.

The cold reading, at its best, provides the client with a character assessment that is uniquely tailored to fit him or her. The reader begins with the same assumptions that guide the psychological reader who relies on the stock spiel. These assumptions are (1) that we all are basically more alike than different; (2) that our problems are generated by the same major transitions of birth, puberty, work, marriage, children, old age, and death; (3) that, with the exception of curiosity seekers and troublemakers, people come to a character reader because they need someone to listen to their conflicts involving love, money, and health.

The cold reader goes beyond these common denominators by gathering as much additional information about the client as possible. Sometimes such information is obtained in advance of the reading. If the reading is through appointment, the reader can use directories and other sources to gather information. When the client enters the consulting room, an assistant can examine the coat left behind (and often the purse as well) for papers, notes, labels, and other such cues about socioeconomic status, and so on. Most cold readers, however do not need such advance information.

The cold reader basically relies on a good memory and acute observation. The client is carefully studied. The clothing- for example, style, neatness, cost, age- provides a host of cues for helping the reader make shrewd guesses about socioeconomic level, conservatism or extroversion, and other characteristics. The client's physical features—weight, posture, looks, eyes, and hands provide further cues. The hands are especially revealing to the good reader. The manner of speech, use of grammar, gestures, and eye contact are also good sources. To the good reader the huge amount of information coming from an initial sizing-up of the client greatly narrows the possible categories into which he classifies clients. His knowledge of actual and statistical data about various subcultures in the population already provides him the basis for making an uncanny and strikingly accurate assessment of the client.

But the skilled reader can go much further in particularizing his reading. He wants to zero in as quickly as possible on the precise problem that is bothering the client. On the basis of his initial assessment he makes some tentative hypotheses. He tests these out by beginning his assessment in general terms, touching upon general categories of problems and watching the reaction of the client. If he is on the wrong track the client's reactions, eye movements, pupillary dilation, other bodily mannerisms—will warn him. When he is on the right track other reactions will tell him so. By watching the client's reactions as he tests out different hypotheses during his spiel, the good reader quickly hits upon what is bothering the customer and begins to adjust the reading to the situation. By this time, the client has usually been persuaded that the reader, by some uncanny means, has gained insights into the client's innermost thoughts. His guard is now down. Often he opens up and actually tells the reader, who is also a good listener, the details of his situation. The reader, after a suitable interval, will usually feed back the information that the client has given him in such a way that the client will be further amazed at how much the reader "knows" about him. Invariably the client leaves the reader without realizing that everything he has been told is simply what he himself has unwittingly revealed to the reader.

## *The Stock Spiel*

The preceding paragraphs indicate that the cold reader is a highly skilled and talented individual. And this is true. But what is amazing about this area of human assessment is how successfully even an unskilled and incompetent reader can persuade a client that he has fathomed the client's true nature. It is probably a tribute to the creativeness of the human mind that a client can, under the right circumstances, make sense out of almost any reading and manage to fit it to his own unique situation. All that is necessary is that the reader make out a plausible case for why the reading ought to fit. The client will do the rest.

You can achieve a surprisingly high degree of success as a character reader even if you merely use a stock spiel which you give to every client. Sundberg (1955), for example, found that if you deliver the following character sketch to a college male, he will usually accept it as a reasonably accurate description of himself: "You are a person who is very normal in his attitudes, behavior and relationships with people. You get along well without effort. People naturally like you, and you are not overly critical of them or yourself. You are neither overly conventional nor overly individualistic. Your prevailing mood is one of optimism and constructive effort, and you are not troubled by periods of depression, psychosomatic illness or nervous symptoms."

Sundberg found that the college female will respond with even more pleasure to the following sketch: "You appear to be a cheerful, well-balanced person. You may have some alternation of happy and unhappy moods, but they are not extreme now. You have few or no problems with your health. You are sociable and mix well with others. You are adaptable to social situations. You tend to be adventurous. Your interests are wide. You are fairly self-confident and usually think clearly."

Sundberg conducted his study over 20 years ago. But the sketches still work well today. Either will tend to work well with both sexes. More recently, several laboratory studies have had excellent success with the following stock spiel (Snyder and Shenkel 1975).

> Some of your aspirations tend to be pretty unrealistic. At times you are extroverted, affable, sociable, while at other times you are introverted, wary and resented. You have found it unwise to be too frank in revealing yourself to others. You pride yourself on being an independent thinker and do nor accept others' opinions without satisfactory proof. You prefer a certain amount of change and variety and become dissatisfied when hemmed in by restrictions and limitations. At times you have serious doubts as to whether you have

made the right decision or done the right thing. Disciplined and controlled on the outside, you tend to be worrisome and insecure on the inside.

Your sexual adjustment has presented some problems for you. While you have some personality weaknesses, you are generally able to compensate for them. You have a great deal of unused capacity which you have not turned to your advantage. You have a tendency to be critical of yourself. You have a strong need for other people to like you and for them to admire you.

Interestingly enough, the statements in this stock spiel were first used in 1943 by Bertram Forer (1948) in a classroom demonstration of personal validation. He obtained most of them from a newsstand astrology book. Forer's students, who thought the sketch was uniquely intended for them as a result of a personality test, gave the sketch an average rating of 4.26 on a scale of O (poor) to 5 (perfect). As many as 16 our of his 39 students (41 percent) rated it as a perfect fit to their personality. Only five gave it a rating below 4 (the worst being a rating of 2, meaning "average"). Almost 30 years later students give the same sketch an almost identical rating as a unique description of themselves.

## The Technique in Action

The acceptability of the stock spiel depends upon the method and circumstances of its delivery. As we shall later see, laboratory studies have isolated many of the factors that contribute to persuading clients that the sketch is a unique description of themselves. A great deal of the success of the spiel depends upon "setting the stage." The reader tries to persuade the client that the sketch is tailored especially for him or her. The reader also creates the impression that it is based on a reliable and proven assessment procedure. The way the sketch is delivered and dramatized also helps. And many of the rules that I give for the cold reading also apply to the delivery of the stock spiel.

The stock spiel, when properly delivered, can be quite effective. In fact, with the right combination of circumstances the stock spiel is often accepted as a perfect and unique description by the client. But, in general, one can achieve even greater success as a character analyst if one uses the more flexible technique of the cold reader. In this method one plays a sort of detective role in which one takes on the role of a Sherlock Holmes. (See the "Case of the Cardboard Box" for an excellent example of cold reading.) One observes the jewelry, prices the clothing, evaluates the speech mannerisms, and studies the reactions of the subject. Then whatever information these observations provide is pieced together into a character reading which is aimed more specifically at the particular client.

A good illustration of the cold reader in action occurs in a story told by the well-known magician John Mulholland. The incident took place in the 1930s. A young lady in her late twenties or early thirties visited a character reader. She was wearing expensive jewelry, a wedding band, and a black dress of cheap material. The observant reader noted that she was wearing shoes which were currently being advertised for people with foot trouble. (Pause at this point and imagine that you are the reader; see what you would make of these clues.)

By means of just these observations the reader proceeded to amaze his client with his insights. He assumed that this client came to see him, as did most of his female customers, because of a love or financial problem. The black dress and the wedding band led him to reason that her husband had died recently. The expensive jewelry suggested that she had been financially comfortable during marriage, but the cheap dress indicated that her husband's death had left her penniless. The therapeutic shoes signified that she was now standing on her feet more than she was used to, implying that she was working to support herself since her husband's death.

The reader's shrewdness led him to the following conclusion, which turned out to be correct: The lady had met a man who had proposed to her. She wanted to marry the man to end her economic hardship. But she felt guilty about marrying so soon after her husband's death. The reader told her what she had come to hear— that it was all right to marry without further delay.

## The Rules of the Game

Whether you prefer to use the formula reading or to employ the more flexible technique of the cold reader, the following bits of advice will help to contribute to your success as a character reader.

*1. Remember that the key ingredient of a successful character reading is confidence.* If you look and act as if you believe in what you are doing, you will be able to sell even a bad reading to most of your subjects.

The laboratory studies support this rule. Many readings are accepted as accurate because the statements do fit most people. But even readings that would ordinarily be rejected as inaccurate will be accepted if the reader is viewed as a person with prestige or as someone who knows what he is doing.

One danger of playing the role of reader is that you will persuade yourself that you really are divining true character. This happened to me. I starred reading palms when I was in my teens as a way to supplement my income from doing magic and mental shows. When I started I did nor believe in palmistry. But I knew that to "sell" it I had to act as if I did. After a few years I became a firm believer in palmistry. One day the late Dr. Stanley Saks, who was a professional mentalist and a man I respected, tactfully suggested that it would make an interesting experiment if I deliberately gave readings opposite to what the lines indicated. I tried this out with a few clients. To my surprise and horror my readings were just as successful as ever. Ever since then I have been interested in the powerful forces that convince us, reader and client alike, that something is so when it really isn't.

2. *Make creative use of the latest statistical abstracts, polls, and surveys.* This can provide you with a wealth of material about what various subclasses of our society believe, do, want, worry about, and so on. For example, if you can ascertain about a client such things as the part of the country he comes from, the size of the city he was brought up in, his parents' religion and vocations, his educational level and age, you already are in possession of information that should enable you to predict with high probability his voting preferences, his beliefs on many issues, and other traits.

3. *Set the stage for your reading.* Profess a modesty about your talents. Make no excessive claims. This catches your subject off guard. You are not challenging him to a battle of wits. You can read his character; whether he cares to believe you or not is his concern.

4. *Gain his cooperation in advance.* Emphasize that the success of the reading depends as much upon his sincere cooperation as upon your efforts. (After all, you imply, you already have a successful career at reading characters. You are not on trial—he is.) State that due to difficulties of language and communication, you may not always convey the exact meaning which you intend. In these cases he is to strive to reinterpret the message in terms of his own vocabulary and life.

You accomplish two invaluable ends with this dodge. You have an alibi in case the reading doesn't click; it's his fault not yours! And your subject will strive to fit your generalities to his specific life occurrences. Later, when he recalls the reading he will recall it in terms of specifics; thus you gain credit for much more than you actually said.

Of all the pieces of advice this is the most crucial. To the extent that the client is made an active participant in the reading the reading will succeed. The good reader, deliberately or unwittingly, is the one who forces the client to actively search his memory to make sense of the reader's statements.

5. *Use a gimmick such as a crystal ball, tarot cards, or palm reading.* The use of palmistry, say, serves two useful purposes. It lends an air of novelty to the reading: but, more important, it serves as a cover for you to stall and to formulate your next statement. While you are trying to think of something to say next, you are apparently carefully studying a new wrinkle or line in the hand. Holding hands, in addition to any emotional thrills you may give or receive thereby, is another good way of detecting the reactions of the subject to what you are saying (the principle is the same as "muscle reading").

It helps, in the case of palmistry or other gimmicks, to study some manuals so that you know roughly what the various diagnostic signs are supposed to mean. A clever way of using such gimmicks to pin down a client's problem is to use a variant of "Twenty Questions," somewhat like this: Tell the client you have only a limited amount of time for the reading. You could focus on the heart line, which deals with emotional entanglements; on the fate line, which deals with vocational pursuits and money matters; the head line, which deals with personal problems; the health line, and so on. Ask him or her which one to focus on first. This quickly pins down the major category of problem on the client's mind.

6. *Have a list of stock phrases at the tip of your tongue.* Even if you are doing a cold reading, the liberal sprinkling of stock phrases amidst your regular reading will add body to the reading and will fill in time as you try to formulate more precise characterizations. You can use the statements in the preceding stock spiels as a start. Memorize a few of them before undertaking your initial ventures into character reading. Palmistry, tarot, and other fortune telling manuals also are rich sources for good phrases.

7. *Keep your eyes open.* Also use your other senses. We have seen how to size up the client on the basis of clothing, jewelry, mannerisms, and speech.. Even a crude classification on such a basis can provide sufficient information for a good reading. Watch the impact of your statements upon the subject. Very quickly you will learn when you are "hitting home" and when you are "missing the boat."

8. *Use the technique of "fishing."* This is simply a device for getting the subject to tell you about himself. Then you rephrase what he has told you into a coherent sketch and feed it back to him. One version of fishing is to phrase each statement in the form of a question. Then wait for the subject to reply (or react). If the reaction is positive, then the reader turns the statement into a positive assertion. Often the subject will respond by answering the implied question and then some. Later he will tend to forget that he was the source of your information. By making your statements into questions you also force the subject to search through his memory to retrieve specific instances to fit your general statement.

9. *Learn to be a good listener.* During the course of a reading your client will be bursting to talk about incidents that are brought up. The good reader allows the client to talk at will. On one occasion I observed a tea-leaf reader. The client actually spent 75 percent of the total time talking. Afterward when I questioned the client about the reading she vehemently insisted that she had not uttered a single word during the course of the reading. The client praised the reader for having so astutely told her what in fact she herself had spoken.

Another value of listening is that these clients who seek the services of a reader actually want someone to listen to their problems. In addition many clients have already made up their minds about what choices they are going to make. They merely want support to carry out their decision.

10. *Dramatize your reading.* Give back what little information you do have or pick up a little bit at a time. Make it seem more than it is. Build word pictures around each divulgence. Don't be afraid of hamming it up.

11. *Always give the impression that you know more than you are saying.* The successful reader, like the family doctor, always acts as if he knows much more. Once you persuade the client that you know one item of information about him that you could not possibly have obtained through normal channels, the

client will automatically assume you know all. At this point he will typically open up and confide in you.

*12. Don't be afraid to flatter your subject every chance you get.* An occasional subject will protest such flattery, but will still cherish it. In such cases you can further flatter him by saying, "You are always suspect of people who flatter you. You just can't believe that someone will say good of you unless he is trying to achieve some ulterior goal."

*13. Finally remember the golden rule: Tell the client what he wants to hear.*

Sigmund Freud once made an astute observation. He had a client who had been to a fortune teller many years previously. The fortune teller had predicted that she would have twins. Actually she never had children. Yet, despite the fact that the reader had been wrong, the client still spoke of her in glowing terms. Freud tried to figure out why this was so. He finally concluded that at the time of the original reading the client wanted desperately to have children The fortune teller sensed this and told her what she wanted to hear. From this Freud inferred that the successful fortune teller is one who predicts what the client secretly wishes to happen rather than what actually will happen (Freud, 1933).

## The Fallacy of Personal Validation

As we have seen, clients will readily accept stock spiels such as those I have presented as unique descriptions of themselves. Many laboratory experiments have demonstrated this effect. Forer (1948) called the tendency to accept as valid a personality sketch on the basis of the client's willingness to accept it 'the fallacy of personal validation."

The early studies on personal validation were simply demonstrations to show that students, personnel directors, and others can readily be persuaded to accept a fake sketch as a valid description of themselves. A few studies tried to go beyond the demonstration and tease out factors that influence the acceptability of the fake sketch. Sundberg (1955), for example, gave the Minnesota Multiphasic Personality Inventory (known as the MMPI) to 44 students. The MMPI is the most carefully standardized personality inventory in the psychologist's tool kit. Two psychologists, highly experienced in interpreting the outcome of the MMPI, wrote a personality sketch for each student on the basis of his or her test results. Each student then received two personality sketches—the one actually written for him or her—and a fake sketch. When asked to pick which sketch described him or her better, 26 of the 44 students (59 percent) picked the fake sketch!

Sundberg's study highlights one of the difficulties in this area. A fake, universal sketch can be seen as a better description of oneself than can a uniquely tailored description by trained psychologists based upon one of the best assessment devices we have. This makes personal validation a completely useless procedure. But it makes the life of the character reader and the pseudo psychologist all the easier. His general and universal statements have more persuasive appeal than do the best and most appropriate descriptions that the trained psychologist can come up with.

Some experiments that my students and I conducted during the 1950s also supplied some more information about the acceptability of such sketches. In one experiment we gave some students a fake sketch (the third stock spiel previously discussed) and told half of them that it was the result of an astrological reading and the other half that it was the result of a new test, the Harvard Basic Personality Profile. In those days, unlike today, students had a low opinion of astrology. All the students rated each of the individual statements as generally true of themselves. The groups did not differ in their ratings of the acceptability of the individual statements. But when asked to rate the sketch as a whole, the group that thought it came from an accepted personality test rated the acceptability significantly higher than did the group that thought it came from an astrologer. From talking to individual students it was clear that those who were in the personality test group believed that they had received a highly accurate and unique characterization of themselves. Those in the astrology group admitted that the individual statements were applicable to themselves but dismissed the apparent success of the astrology as due to the fact that the statements were so general that they would fit anyone. In other words, by changing the context in which they got the statements we were able to manipulate the subjects' perceptions as to whether the statements were generalities that applied to everyone or were specific characterizations of themselves.

In a further experiment we obtained a pool of items that 80 percent or more of Harvard students endorsed as true of themselves. We then had another group of Harvard students rate these items as "desirable" or "undesirable" and as "general" or "particular" (true of only a few students). Thus we had a set of items that we knew almost all our subjects would endorse as true of themselves, but which varied on desirability and on perceived generality. We were then able to compose fake sketches which varied in their proportion of desirable and specific items. We found that the best recipe for creating acceptable stock spiels was to include about 75 percent desirable items, but ones which were seen as specific, and about 25 percent undesirable items, but ones which were seen as general. The undesirable items had the apparent effect of making the spiel plausible. The fact that the items were seen as being generally true of other students made them more acceptable.

The most extensive program of research to study the factors making for acceptability of fake sketches is that by C. R Snyder and his associates at the University of Kansas. A brief summary of many of his findings was given in an article in *Psychology Today* (Snyder and Shenkel 1975). In most of his studies Snyder uses a control condition in which the subject is given the fake sketch and told that this sketch is generally true for all people. On a rating scale from 1 to 5 (1, very poor; 2, poor; 3, average; 4, good; 5, excellent) the subject rates how well the interpretation fits his personality. A typical result for this control condition is a rating of around 3 to 4, or between average and good. But when the sketch is presented to the subject as one which was written "for you, personally" the acceptability tends to go up to around 4.5, or between good and excellent.

In a related experiment the subjects were given the fake sketch under the pretense that it was based on an astrological reading. The control group, given the sketch as "generally true for all people," rated it about 3.2, or just about average. A second group was asked to supply the astrologer with information on the year and month of their birth. When they received their sketches they rated them on the average at 3.76, or just below good. A third group supplied the mythical astrologer with information on year, month, and day of birth. These subjects gave a mean rating of 4.98.

From experiments such as those we have learned the following. The acceptability of a general sketch is enhanced when (1) the reader or source is believed to know what he is doing. (2) the instrument or assessment device is plausible, (3) a lot of mumbo jumbo is associated with the procedure (such as giving month, day, hour, and minute of birth along with a lot of complicated calculations), and (4) the client is led to believe that the sketch has been tailored to his personality. When these conditions are met, the client, and possibly the reader as well, have a strong "illusion of uniqueness"—that is, the client is persuaded that the sketch describes himself or herself and no one else.

## Why Does it Work?

But why does it work? And why does it work so well? It does not help to say that people are gullible or suggestible. Nor can we dismiss it by implying that some individuals are just not sufficiently discriminating or lack sufficient intelligence to see through it. Indeed one can argue that it requires a certain degree of intelligence on the part of a client for the reading to work well. Once the client is actively engaged in trying to make sense of the series of sometimes contradictory statements issuing from the reader, he becomes a creative problem-solver trying to find coherence and meaning in the total set of statements. The task is not unlike that of trying to make sense of a work of art, a poem, or, for that matter, a sentence. The work of art, the poem, or the sentence serve as a blueprint or plan from which we can construct a meaningful experience by bringing to bear our own past experiences and memories.

In other words the reading succeeds just because it calls upon the normal processes of comprehension that we ordinarily bring to bear in making sense our of any form of communication. The raw information in a communication is rarely, if ever, sufficient in itself for comprehension. A shared context and background is assumed. Much has to be filled in by inference. The good reader, like anyone who manipulates our perceptions, is merely exploiting the normal processes by which we make sense out of the disorderly array of inputs that constantly bombard us.

A variety of examples and experiments could be drawn upon to illustrate this point. Let us look at a couple of instructive cases. Dr. Joseph Weizenbaum, of M.I.T., developed a computer program that would enable a human to apparently carry on a conversation with a computer. One version of his program is called "Doctor" The program is designed to "talk" like a Rogerian psychotherapist. Such a therapist is relatively easy to imitate because the basis of Rogers' technique is simply to feed the patient's statements back to him in a paraphrased form. The human, playing the role of the patient, types in questions and answers on a teletype machine. The computer, under control of the program, types replies and questions on the same teletype paper. A typical example (Weizenbaum, 1976) of a conversation between a young lady and Doctor follows. (The lady speaks first. The computer's responses are printed in italics.)

Men are all alike.
*In what way?*
They're always bugging us about something or other.
*Can you think of a specific example?*
Well, my boyfriend made me come here.
*Your boyfriend made you come here?*
He says I'm depressed much of the time.
*I am sorry to hear you are depressed.*
It's true. I am unhappy.
*Do you think coming here will help you not to be unhappy?*
I need some help, that much seems certain.
*What would it mean to you if you got some help?*
Perhaps I could learn to get along with my mother.
*Tell me more about your family.*
My mother takes care of me.
*Who else in your family takes care of you?*
My father.
*Your father?*
You are like my father in some ways.
*What resemblance do you see?*
You are not very aggressive, but I think you don't want me to notice that.
*What makes you think I am not very aggressive?*
You don't argue with me.
*Why do you think I don't argue with you?*
You are afraid of me.
*Does it please you to believe I am afraid of you?*
My father is afraid of everybody
*What else comes to mind when you think of your father?*
Bullies.
*Does that have anything to do with the fact that your boyfriend made you come here?*

Doctor is a quite primitive program as natural-language programs go. It employs a lot of tricks and stock phrases. It has no mechanisms for actually understanding sentences. Instead it seeks out keywords that are typed and does some simple syntactical transformations. For example, if the program sees a sentence of the form "Do you X" it automatically prints out the response "What makes you think I X?" When Doctor cannot match the syntax of a given sentence it can cover up in two ways. It can say something noncommittal, such as *"Please go on"* or *"What does that suggest to you?"* Or it can recall an earlier match and refer back to it, as for example, *"How does this relate to your depression?"* where depression was an earlier topic of conversation.

In essence Doctor is a primitive cold reader. It uses stock phrases to cover up when it cannot deal with a given question or input. And it uses the patient's own input to feed back information and create the illusion that it understands and even sympathizes with the patient. This illusion is so powerful that patients, even when told they are dealing with a relatively simple-minded program, become emotionally involved in the interaction. Many refuse to believe that they are dealing with a program and insist that a sympathetic human must be at the control at the other end of the teletype.

Sociologist Harold Garfinkel has supplied another instructive example. He conducted the following experiment. The subjects were told that the Department of Psychiatry was exploring alternative means to therapy "as a way of giving persons advice about their personal problems." Each subject was then asked to discuss the background of some serious problem on which he would like advice. After having done this the subject was to address some questions which could be answered "yes or "no" to the "counselor" (actually an experimenter). The experimenter-counselor heard the questions from an adjoining room and supplied a "yes" or "no answer to each question after a suitable pause. Unknown to the subject, the series of yes-no answers had been pre-programmed according to a table of random numbers and was not related to his questions. Yet the typical subject was sure that the counselor fully understood the subject's problem and was giving him sound and helpful advice.

Let me emphasize again that statements as such have no meaning. They convey meaning only in context and only when the listener or reader can bring to bear his large store of worldly knowledge. Clients are not necessarily acting irrationally when they find meaning in the stock spiels or cold reading. Meaning is an interaction of expectations, context, memory, and given statements. An experiment by the Gestalt psychologist Solomon Asch (1948) will help make this point. Subjects were given the following passage and asked to think about it: "I hold it that a little rebellion, now and then, is a good thing, and as necessary in the political world as storms are in the physical." One group of subjects was told that the author of the passage was Thomas Jefferson (which happens to be true). The subjects were asked if they agreed with the passage and what it meant to them. These subjects generally approved of it and interpreted the word *rebellion* to mean minor agitation. Bur when subjects were given the same passage and told that its author was Lenin, they disagreed with it and interpreted *rebellion* to mean a violent revolution.

According to some social psychologists the different reactions show the irrationality of prejudice. But Asch points out that the subjects could be acting quite rationally. Given what they know about Thomas Jefferson and Lenin, or what they believe about them, it makes sense to attribute different meanings to the same words spoken by each of them if one thinks that Jefferson believed in orderly government and peaceful processes, then it would not make sense to interpret his statement to actually mean a bloody or physical revolution. If one thinks that Lenin favors war and bloodshed, then it makes sense, when the statement is attributed to him, to interpret *rebellion* in its more extreme term.

Some recent research that my colleagues and I conducted might also be relevant here. Our subjects were given the task of forming an impression of a hypothetical individual on the basis of a brief personality sketch. In one condition the subjects were given a sketch that generally led to an impression of a nice, personable, friendly sort of fellow. In a second condition the subjects were given a sketch that created an impression of a withdrawn, niggardly [meaning miserly or stingy] individual. Both groups of subjects were then given a new sketch that supposedly contained more information about the hypothetical individual. In both cases the subjects were given an identical sketch. This sketch contained some descriptors that were consistent with the friendly image and some that were consistent with the niggardly image. The subjects were later tested to see how well they recognized the actual adjectives that were used in the second sketch. One of the adjectives, for example, was charitable. The test contained foils for each adjective. For example, the word *generous* also appeared on the test but did not appear in the sketch. Yet subjects who had been given the friendly impression checked *generous* just as frequently as they checked charitable. But subjects in the other condition did not confuse *charitable* with *generous*. Why? Because, we theorize, the two different contexts into which charitable had to be integrated produced quite different meanings. When subjects who have already built up an impression of a "friendly" individual encounter the additional descriptor *charitable*, it is treated as merely further confirmation of their general impression. In that context *charitable* is simply further confirmation of the nice-guy image. Consequently when these subjects are asked to remember what was actually said they can remember only that the individual was further described in some way to enhance the good-guy image, and *generous* is just as good a candidate for the description as is charitable in that context.

But when the subjects who have an image of the person as a withdrawn, niggardly individual encounter *charitable*, the last thing that comes to mind is generosity. Instead, they probably interpret charitable as implying that he donates money to charities as a way of gaining tax deductions. In this latter condition the subjects have no subsequent tendency to confuse *charitable* with *generous*.

The cold reading works so well, then, because it taps a fundamental and necessary human process. We have to bring our knowledge and expectations to bear in order to comprehend anything in our world. In most ordinary situations this use of context and memory enables us to correctly interpret statements and supply the necessary inference to do this. But this powerful mechanism can go astray in situations where there is no actual message being conveyed. Instead of picking up random noise we still manage to find meaning in the situation. So the same system that enables us to creatively find meanings and make new discoveries also makes us extremely vulnerable to exploitation by all sorts of manipulators. In the case of the cold reading the manipulator may be conscious of his deception; but often he, too, is a victim of personal validation.

## REFERENCES

**Originally published in *The Zetetic* (Spring/Summer 1977): 18-37.**

Asch, S. E. 1948. "The Doctrine of Suggestion, Prestige, and Imitations in Social Psychology." *Psychological Review,* 55: 250–76.

Forer, B. R. 1949. "The Fallacy of Personal Validation: A Classroom Demonstration of Gullibility." *Journal of Abnormal and Social Psychology,* 44: 118–23.

Freud, S. 1933. *New Introductory Lectures on Psychoanalysis.* New York: W. W. Norton.

Garfinkel, H. 1967. *Studies in Ethnomethodology.* Englewood-Cliffs, N.J.: Prentice-Hall.

Snyder, C. R., and R. J. Shenkel 1975. "The P. T. Barnum Effect." *Psychology Today* 8: 52–54.

Sundberg, N. D. 1955. "The Acceptability of 'Fake' versus 'Bona Fide' Personality Test Interpretations." *Journal of Abnormal and Social Psychology,* 50: 145–57.

Weizenbaum, J. 1976. *Computer Power and Human Reason.* San Francisco: Freeman.

# 4

## SYLVIA BROWNE:
## PSYCHIC GURU OR QUACK?

### *Bryan Farha*

Most assuredly you've heard the phrase "innocent until proven guilty." I'm pretty much a believer in that saying. Our legal system is built around it—and justifiably so. But what if an alleged psychic makes three promises on international television to test her extraordinary claims, yet makes no effort to do so? Should the phrase for that person become "guilty until proven innocent?" In any event, I'm referring to one of the most popular "spiritual mediums" in the country—if not the world. Sylvia Browne—herein after referred to simply as "Sylvia."

Sylvia's Web site (www.sylvia.org) states:

> Sylvia began her professional career as a psychic on May 8, 1973 with a small meeting in her home. Within one year her practice had grown very large and she decided to incorporate her business as the Nirvana Foundation for Psychic Research. Wanting to make her work as professional as possible, then as now, Sylvia maintains required business licenses, is a member of a national consumer protection agency, and donates a lot of time to charitable organizations and working with police. Her business has remained in the same general vicinity since beginning her work.
>
> Sylvia's family line includes several practicing psychics and mediums. Her maternal grandmother, Ada, was an established and respected counselor and healer in Kansas City MO. This familial psychic talent has also passed on to her son Christopher Dufresne. There seems to be a genetic component necessary to create exceptional psychics, Sylvia's blood line carries that predisposition to excellence. As Sylvia says, "Anyone can learn to play the piano, but not everyone is a concert pianist."[1]

Sylvia "diagnoses" health problems, purports to communicate with the dead, and even claims to have proven there is an afterlife. Her recent books include *Contacting Your Spirit Guide* and *Past Lives, Future Healing: A Psychic Reveals the Secrets to Good Health and Great Relationships*. For several years, she has been popularized by TV talk-show hosts Montel Williams and Larry King. Montel, who has hosted Sylvia more than 70 times since 1995[2], will have absolutely no part of skeptical perspectives. Larry has included skeptics as guests on two of her three recent appearances during the past three years. But neither Montel nor Larry has shown the slightest interest in checking out her monumental claims—and I do mean monumental. And, as far as I know, neither Montel, Larry, or Sylvia have investigated—or even care—whether Sylvia's health advice causes people to delay appropriate treatment or to undergo needless tests to look for nonexistent problems that Sylvia "sees."

With respect to health, you can get a "reading" from Sylvia for only $700 by phone or $750 in person.[3] Does that sound like proper commercial activity for someone who has no medical license, just a master's degree in English Literature?

## Sylvia in Action

On September 3, 2001, Sylvia was challenged on "Larry King Live" by James Randi, the conjurer/skeptic who heads the James Randi Educational Foundation (JREF), a nonprofit organization founded in 1996 to "promote critical thinking by reaching out to the public and media with reliable information about paranormal and supernatural ideas so widespread in our society today."[4] During the program, Sylvia claimed that she had worked with 350 doctors and had cured a child of seizures. When asked how he thought Browne worked, Randi replied that she asks questions, makes guesses, offers suggestions, throws out words, wait for answers, and builds on what she gets  a method commonly referred to as "cold reading." Randi also pointed out that people tend to remember what seems to fit and forget what does not. During the program, she demonstrated her technique to one caller, as follows:

> **Caller:** Sylvia and Larry, I enjoy you so very, very much. I listened to you for years and I just wanted to get on for a long time. Randi, I feel sorry for you as well, because we have to believe in something. My question is, Sylvia, I never had a chance to say good-bye to my husband. And I am wondering if he knows how much I loved him.
>
> **Browne:** Not only did he know that, but what was the—clot or whatever that let loose? Because it looks like there was something about a clot.
>
> **Caller:** Yes, he had a severe brain hemorrhage at the very last minute.
>
> **Browne:** Because it looks likes it was, not only that, but this was massive.
>
> **Caller:** Yes, it was.

**Browne:** Yeah.

**Caller:** Right through the top of his head.

**Browne:** And he really—you know, there are so many times, like when I lost so many people. I don't care how many times if you can say good-bye, you never have enough good-byes. But see, aside from Randi, he hears everything you say, especially when you talk to him.

**Caller:** Well, I don't really know whether I can say anything to him. There are people like that. But I feel sorry for them. Because we have to believe.

**King:** I thank you, ma'am. (Turning to Sylvia): Now, help me with something.

**Browne:** Yes?

**King:** Did you see that clot?

**Browne:** I saw the clot letting loose.

**King:** How do you explain? . . . .

**Browne:** I don't know. It's like Randi said one time to one of the psychics, a lot of psychics just say chest. Of course, because a lot of people have chest problem. But not everybody has a massive embolism.

**King:** How would you explain that. A massive...

**Browne:** I know what he is going to say, it's a guess.

**Randi:** Larry, you're asking me to explain specific things. I don't know who this woman is who called. I don't know whether she is a ringer. I'm not saying she is, and I'm not suggesting that.

**Browne:** Oh.

**Randi:** But it is possible. There are many possibilities here. We have made a lucky guess, and we have hit. An embolism. A clot.

**King:** There are many possibilities. Is one of them, Randi—is one of the possibilities Sylvia is right.

**Randi:** Absolutely.[5]

Randi was actually overgenerous. A clot might be involved in a heart attack, a stroke, or a few other rapidly fatal conditions. Because heart attacks and strokes are among the most common causes of sudden death, the word "clot" had a fairly good chance of being correct. However, in this case it was not. Although Browne and King seemed to think that Browne's "diagnosis" was on target, it actually was dead wrong. The caller said the problem was a severe brain hemorrhage. A clot is just the opposite of a hemorrhage. As Randi notes on his Web site:

> Now, to me, this sounds as if the caller is describing an impact of some sort to the top of the head! Clots don't go through the top of the head. They originate inside the head and stay there. Notice, too, that the term "embolism," which was introduced by Sylvia as applying to the cause of this death, and never by the caller, refers to a blocked blood vessel, and could not apply here. She said, amplifying her reading, - the caller had already been disconnected by that time - that I claimed "psychics" frequently refer to "chest problems" as a

cause of death, while "not everyone has a massive embolism." She then predicted what I would say about this remarkable "hit," that I would call it a guess. She was wrong; I say that it's a dead miss. And it is. No, not everyone has a massive embolism, nor a clot, both of which Sylvia put forth as the cause of death, and this man had neither.

An MD friend said that in his opinion, Sylvia is not just full of baloney but also dangerous. She mentioned to one caller that she should check the "bilirubin," which she told King "is a liver enzyme." The fact is that bilirubin is not a liver enzyme but a degradation product of human hemoglobin. This is routinely checked when blood tests are done. No need to check it separately, as any elevation of bilirubin will give the very obvious clinical appearance of jaundice. You just have to look in the persons eye to see that. And there is no test for Epstein-Barr disease related to the examination of fecal matter, as Sylvia, in her vast medical expertise, offered to a caller. And, she prescribed the drug Tegretol, as well, for another caller's disorder. This type of medical advice, which by law Sylvia cannot offer, is dangerous as it can mislead the caller. Who is she to give medical advice? Larry King was amazed at her facility with medical terms. Facility does not necessarily equate with accuracy.[6]

Later in the program, Browne said that Randi needed to see a doctor because had a problem in his left ventricle (the chamber in the heart from which the blood is squeezed out into the body's general circulation). Soon afterward, Randi saw a cardiac surgeon, who found no problem. If you think this example is benign, consider that most of Sylvia's readings are with people who believe in her alleged psychic ability and therefore take her seriously. Health-related readings like this are commonplace with Sylvia.

## One Evasion after Another

The James Randi Educational Foundation (JREF) offers a one-million-dollar prize to anyone who can show, under proper observing conditions, evidence of any paranormal, supernatural, or occult power or event. The prize is in the form of negotiable bonds held in a special investment account. All tests are designed with the participation and approval of the applicant. In most cases, the applicant will be asked to perform a relatively simple preliminary test of the claim, which if successful, will be followed by the formal test. Preliminary tests are usually conducted by associates of the JREF at the site where the applicant lives. Upon success in the preliminary testing process, the "applicant" becomes a "claimant." So far, no one has ever passed the preliminary tests.[7]

Sylvia has promised three times to take the test. On March 6, 2001 Larry King Live hosted a discussion about criticism aimed at medium John Edward, who hosts "Crossing Over." Sylvia and another alleged psychic (James Van Praagh) participated together with skeptics Leon Jaroff and Paul Kurtz and three others. During this program, Sylvia insisted that Kurtz had a prostate problem, which Kurtz denied. Jaroff urged Sylvia and Van Praagh to take Randi's "million-dollar challenge" and Sylvia agreed to do so:

> **Browne**: I have never been offered this challenge.
> **King**: You would take it?
> **Browne**: I would take the challenge. I have tried to run around the table [Randi] ran away from me.
> **King**: She will meet with Randi and take the challenge.
> **Browne**: He ran way from me.[8]

Browne failed to contact Randi, but on the September 3 show, she told Randi she would take the test.

> **King**: Randi has offered a million dollars in the past to those who would take his challenge. Would you first—let's start with Randi—explain what the challenge constitutes? You will pay a million dollars if?
> **Randi**: A million dollars in negotiable bonds, Larry, to any person or persons who can provide evidence of any paranormal, occult or supernatural event or ability of any kind under proper observing conditions. It is that simple.
> **King**: OK, and the observing conditions would be?
> **Randi**: It would depend upon what the claim is. I have got a whole outline right here that will tell Sylvia exactly what the test would be if she agrees to take the test.
> **King**: Sylvia, in the past you have not agreed to this.
> **Browne**: Well, I don't even want his million dollars. I don't want his million dollars. I mean, the reason I came on is because he kept you know, my web site, yeah, yeah, and said I would never come on and face him. But I don't care about his million dollars. I mean, I don't need his million. . . .
> **King**: Are you willing to take his test?
> **Browne**: Yeah, whatever test it is.

After Randi suggested the specific type of psychic ability he would test, Sylvia agreed: All that was needed was for her to contact Randi. But by April 2003, she had made no contact. On May 16, she appeared again on "Larry King Live," this time as the only guest. As usual, the program began with Larry's unskeptical questions plus phone calls from viewers who sounded like true believers. About 40 minutes into the show, I managed to get past the screeners by telling them I wanted to ask about "my dead cousin." I'm not proud of being deceptive, but I don't believe the screeners would have let me through if they knew that I would question her about Randi's test. As far as I know, nobody has ever been able to do this while she was on the air. Here's what took place:

> **King**: Oklahoma City, hello.

**Caller (me):** Sylvia, 620 days [ago] on Larry's show, you agreed to take James Randi's $1 million paranormal challenge, and on a later show you even agreed to the specific terms of the test.

**Browne:** Yes, but let me tell you something. I also found out that he won't put it into escrow. He won't put the money into escrow.

**Caller:** You agreed to the terms of the test.

**Browne:** No, not until he puts the money into escrow. I mean, why would I do it when the money can't be validated?

**Caller:** Have you contacted James?

**Browne:** I don't want to contact him. I already know about this Russian person who the lawyer contacted and said he won't put it into escrow.

**Caller:** OK, so you agreed 620 days ago to take his test.

**Browne:** I'm not going to do that—I'm not going to do that if he doesn't have the money.

**Caller:** If I can arrange for James to come up with the money, would you take the test?

**King:** You said you would.

**Browne:** Yes, yes, I will. But if he won't come up with the other girl, why would he come up with me?

**King:** If you come up with it, sir, she'll do it.

**Caller:** And will you arrange for it, Larry?

**King:** Sure.

**Caller:** You got it.

**King:** Be happy to do it.[9]

Promise #3, this time from Sylvia *and* Larry King: Larry will arrange for the testing, and Sylvia will take the test if the money *she previously dismissed as unimportant* can be validated. Randi, who has posted a "Sylvia Browne's Clock" webpage at http://www.randi.org/sylvia/index.shtml, corrected my figures. Sylvia had agreed to take the test 808 days before I had called—620 was the number of days since she had agreed to the specific protocol.

On May 18th, Randi emailed me a scanned copy of document from Goldman, Sachs & Company stating that the JREF prize money account contained $1,054,656.70. I immediately wrote to Larry King, with copies to Randi and Sylvia, and Randi sent the following letter to both by certified mail:

Ms. Browne:

Though proof of the JREF prize money is easily available on request, you have not made any such request. Your May 16th appearance on the Larry King Live TV show, seemed to indicate that you were ignorant of the facts, and since we are an educational foundation, we therefore enclose a notarized copy of the account status showing the balance in a special "James Randi Educational Foundation Prize Account" in excess of one million dollars. Also enclosed is a formal statement from the agency holding these assets, verifying that the funds are in place. I'm sure that you are aware of the grave legal consequences that would result against the JREF, if either of these documents were to be found false or altered.

As you are also aware, we have legally committed ourselves to awarding this prize money to anyone who successfully passes both the preliminary and then the formal test, as agreed to between the applicant and the JREF. This is described on our web page, which also clearly states all the conditions for assuring that the prize money will be awarded if the conditions are met. Since you have already heard and accepted the terms and protocol of the test, and your understanding and agreement have been broadcast across the world via CNN, it only remains for you to give us a date upon which we can conduct the test.

One caveat: Several of the persons who responded more than a year ago to our request for suitable subjects—one of which would be chosen at random—have since died. It would be necessary for us to re-issue the request, of course, and that would mean that a suitable date would have to be set sometime in July, but no sooner.

Now that this issue of the prize money has been resolved, and there can no longer be any impediment to your involvement, we anticipate hearing from you with a renewed acceptance of our challenge. Of course, if you are afraid of taking the test, or you are aware that you cannot pass a simple double-blind test of your claims, you may wish to further obfuscate the matter by producing more excuses and problems. That's entirely up to you.

Since Larry King has agreed to "arrange" that you be assured of the existence and availability of the prize money, a copy of this letter is being sent to him for his information.[10]

On May 22, Sylvia refused to accept Randi's letter.[11] On May 26th, I emailed Sylvia a copy of Randi's letter and asked "any office personnel" who receive it to make sure she reads it herself. On May 27th, I left a telephone message for Larry King's producers, to which they have not responded.

## The Bottom Line

Sylvia Browne would like people to believe she has the psychic ability to communicate with the dead and to diagnose their ailments. She has broken three promises made on international television to take the JREF One Million Dollar Paranormal Challenge. More than two years have passed since her first promise. I don't believe she ever intended to take the test. Do you think any talk-show hosts will care?

## NOTES

**Originally published at www.quackwatch.org.**

1. "Brief History" at http://www.sylvia.org. Accessed, June 30, 2003.

2. Burelle's Information Services at http://www.burrelles.com/transcripts/syndicated/mwil.htm offers transcripts to 67 "Montel" appearances by Sylvia from the fourth quarter of 1995 through the fourth quarter of 2002. A few quarters do not have complete lists; and transcripts for 2003 have not been posted.

3. "Psychic Readings" at http://www.sylvia.org/home/readings.cfm. Accessed, June 30, 2003.

4. "About the Foundation" at http://www.randi.org/jref/index.html, accessed June 30, 2003.

5. "Are Psychics Real?" at http://www.cnn.com/TRANSCRIPTS/0109/03/lkl.00.html, Sept 3, 2001.

6. Randi J. http://www.randi.org/jr/090701.html. *Swift*, Sept 7, 2001.

7. Randi J. "One Million Dollar Paranormal Challenge," at http://www.randi.org/research/index.html, accessed June 30, 2003.

8. Interview with Sylvia Browne, http://www.cnn.com/TRANSCRIPTS/0305/16/lkl.00.html, May 16, 2003.

9. "Are psychics for real?" at http://www.cnn.com/TRANSCRIPTS/0103/06/lkl.00.html, March 6, 2001.

10. Randi J. http://www.randi.org/jr/052303.html, *Swift*, May 23, 2003.

11. http://www.usps.com/shipping/trackandconfirm.htm. Enter 7003 0500 0002 3034 8133 into Track & Confirm" function.

# 5

## DECONSTRUCTING THE DEAD

### *Michael Shermer*

History is not just one damn thing after another, it is also the same damn thing over and over—time's arrow and time's cycle. Fads come and go, in clothing, cars, and psychics. In the 1970s it was Uri Geller, in the 1980s it was Shirley MacLaine, in the 1990s it was James Van Praagh, and to kick off the new millennium it is John Edward. Edward's star is rising rapidly with a hit daily television series "Crossing Over" on the Sci Fi network and a New York Times bestselling book *One Last Time*. He has appeared, unopposed, on *Larry King Live* and has been featured on *Dateline, Entertainment Tonight*, and an HBO special.

Last year, *Skeptic* magazine was the first national publication to run an expose of John Edward in James "The Amazing" Randi's column (in Vol. 8, No. 3, available at www.skeptic.com), a story that was picked up this week by *Time* magazine, who featured a full-page article on what is rapidly becoming the Edward phenomenon. There is, in reality, nothing new here. Same story, different names. In watching Edward I'm amazed at how blatant he is in stealing lines from medium James Van Praagh. It reminds me of entertainers, comedians, and magicians who go to each others' shows to glean new ideas.

*Time*'s reporter Leon Jaroff, quoting from the *Skeptic* article, wrote a skeptical piece in which he reported the experiences of an audience member from an Edward taping. His name is Michael O'Neill, a New York City marketing manager, who reported his experiences as follows (quoting from the *Skeptic* article):

> I was on the John Edward show. He even had a multiple guess "hit" on me that was featured on the show. However, it was edited so that my answer to another question was edited in after one of his questions. In other words, his question and my answer were deliberately mismatched. Only a fraction of what went on in the studio was actually seen in the final 30 minute show. He was wrong about a lot and was very aggressive when somebody failed to

acknowledge something he said. Also, his "production assistants" were always around while we waited to get into the studio. They told us to keep very quiet, and they overheard a lot. I think that the whole place is bugged somehow. Also, once in the studio we had to wait around for almost two hours before the show began. Throughout that time everybody was talking about what dead relative of theirs might pop up. Remember that all this occurred under microphones and with cameras already set up. My guess is that he was backstage listening and looking at us all and noting certain readings. When he finally appeared, he looked at the audience as if he were trying to spot people he recognized. He also had ringers in the audience. I can tell because about fifteen people arrived in a chartered van, and once inside they did not sit together.

Later, an ABC television producer flew out from New York to film me for an investigation of Edward they are conducting. The segment began as a "puff piece" (as she called it), but a chance encounter in the ABC cafeteria with *20/20* correspondent Bill Ritter, with whom I worked on an expose of medium James Van Praagh a few years ago, tipped her off that Edward was, in fact, a Van Praagh clone and that his talking to the dead was nothing more than the old magicians' cold reading trick. After watching the *20/20* piece the producer immediately realized what was really going on inside Edward's studio. She began to ask a few probing questions and was promptly cut off by Edward and his producers. ABC was told they would not be allowed to film inside the studio and that they, the Sci-Fi network, would provide edited clips that ABC could use. The ABC producer became suspicious, and then skeptical. She has been trying to get an interview with Edward to confront him with my critiques, but they continue to put her off. In fact, she later phoned to tell me that Edward's publicist just left a message on her voice mail (with a date and time) stating that Edward was not available for an interview because he is out of state, yet the producer just caught him on television live in studio on CBS New York! Something fishy is going on here and I know what it is.

The video clips I was shown makes it obvious why Edward does not want raw footage going out to the public—he's not all that good at doing cold readings. Where I estimated Van Praagh's hit rate at between 20–30 percent, Edward's hit rate at between 10–20 percent (the error-range in the estimates is created by the fuzziness of what constitutes a "hit"—more on this in a moment). The advantage Edward has over Van Praagh is his verbal alacrity. Van Praagh is Ferrari fast, but Edward is driving an Indy-500 racer. In the opening minute of the first reading captured on film by the ABC camera, I counted over one statement per second (ABC was allowed to film in the control room under the guise of filming the hardworking staff, and instead filmed Edward on the monitor in the raw). Think about that—in one minute Edward riffles through 60 names, dates, colors, diseases, conditions, situations, relatives, and the like. It goes so fast that you have to stop tape, rewind, and go back to catch them all. When he does come up for air the studio audience members to whom he is speaking look like deer in the headlights. In the edited tape provided by Edward we caught a number of editing mistakes, where he appears to be starting a reading on someone but makes reference to something they said "earlier." Oops!

Edward begins by selecting a section of the studio audience of about 20 people, saying things like "I'm getting a George over here. I don't know what this means. George could be someone who passed over, he could be someone here, he could be someone that you know," etc. Of course such generalizations lead to a "hit" where someone indeed knows a George, or is related to a George, or is a George. Now that he's targeted his mark, the real reading begins in which Edward employs cold reading, warm reading, and hot reading techniques.

*1. Cold Reading.* The first thing to know is that John Edward, like all other psychic mediums, does not do the reading—his subjects do. He asks them questions and they give him answers. "I'm getting a P name. Who is this please?" "He's showing me something red. What is this please?" And so on. This is what is known in the mentalism trade as cold reading, where you literally "read" someone "cold," knowing nothing about them. You ask lots of questions and make numerous statements, some general and some specific, and sees what sticks. Most of the time Edward is wrong. If the subjects have time they visibly nod their heads "no." But Edward is so fast that they usually only have the time or impetus to acknowledge the hits. And Edward only needs an occasional strike to convince his clientele he is genuine.

*2. Warm Reading.* This is utilizing known principles of psychology that apply to nearly everyone. For example, most grieving people will wear a piece of jewelry that has a connection to their loved one. Katie Couric on The Today Show, for example, after her husband died, wore his ring on a necklace when she returned to the show. Edward knows this about mourning people and will say something like "do you have a ring or a piece of jewelry on you, please?" His subject cannot believe her ears and nods enthusiastically in the affirmative. He says "thank you," and moves on as if he had just divined this from heaven. Most people also keep a photograph of their loved one either on them or near their bed, and Edward will take credit for this specific hit that actually applies to most people.

Edward is facile at determining the cause of death by focusing either on the chest or head areas, and then exploring whether it was a slow or sudden end. He works his way down through these possibilities as if he were following a computer flow chart and then fills in the blanks. "I'm feeling a pain in the chest." If he gets a positive nod, he continues. "Did he have cancer, please? Because I'm seeing a slow death here." If he gets the nod, he takes the hit. If the subject hesitates at all, he will quickly shift to heart attack. If it is the head, he goes for stroke or head injury from an automobile accident or fall. Statistically speaking there are only half a dozen ways most of us die, so with just a little probing, and the verbal and nonverbal cues of his subject, he can appear to get far more hits than he is really getting.

*3. Hot Reading.* Sometimes psychic mediums cheat by obtaining information on a subject ahead of time. I do not know if Edward does research or uses shills in the audience to get information on people, or even plants in the audience on which to do readings, but in my investigation of James Van Praagh I discovered from numerous television producers that he consciously and deliberately pumps them for information about his subjects ahead of time, then uses that information to deceive the viewing public that he got it from heaven.

The ABC producer also asked me to do a reading on her. "You know absolutely nothing about me so let's see how well this works." After reviewing the Edward tapes I did a ten minute reading on her. She sat there dropped jawed and wide eyed, counting my hits. She proclaimed that I was unbelievably accurate. How did I do it? Let's just say I utilized all three of the above techniques.

Most of the time, however, mediums do not need to cheat. The reason has to do with the psychology of belief. This stuff works because the people who go to mediums want it to work (remember, they do the readings, not the mediums). The simplest explanation for how mediums can get away with such an outrageous claim as the ability to talk to the dead is that they are dealing with a subject the likes of which it would be hard to top for tragedy and finality—death. Sooner or later we all will face this inevitability, starting, in the normal course of events, with the loss of our parents, then siblings and friends, and eventually ourselves. It is a grim outcome under the best of circumstances, made all the worse when death comes early or accidentally to those whose "time was not up." As those who traffic in the business of loss, death, and grief know all too well, we are often at our most vulnerable at such times. Giving deep thought to this reality can cause the most controlled and rational among us to succumb to our emotions.

The reason John Edward, James Van Praagh, and the other so-called mediums are unethical and dangerous is that they are not helping anyone in what they are doing. They are simply preying on the emotions of grieving people. As all loss, death, and grief counselors know, the best way to deal with death is to face it head on. Death is a part of life, and pretending that the dead are gathering in a television studio in New York to talk twaddle with a former ballroom-dance instructor is an insult to the intelligence and humanity of the living.

## NOTE

Originally published as "Talking Twaddle with the Dead: The Tragedy of Death—the Farce of James Van Praagh," *Skeptic* 6, no. 1 (1998): 48–53.

# 6

## NEAR-DEATH EXPERIENCES:
## IN OR OUT OF BODY?

### Susan Blackmore

What is it like to die? Although most of us fear death to a greater or lesser extent, there are now more and more people who have "come back" from states close to death and have told stories of usually very pleasant and even joyful experiences at death's door.

For many experiencers, their adventures seem unquestionably to provide evidence for life after death, and the profound effects the experience can have on them is just added confirmation. By contrast, for many scientists these experiences are just hallucinations produced by the dying brain and of no more interest than an especially vivid dream.

So which is right? Are near-death experiences (NDEs) the prelude to our life after death or the very last experience we have before oblivion? I shall argue that neither is quite right: NDEs provide no evidence for life after death, and we can best understand them by looking at neurochemistry, physiology, and psychology; but they are much more interesting than any dream. They seem completely real and can transform people's lives. Any satisfactory theory has to understand that too—and that leads us to questions about minds, selves, and the nature of consciousness.

### Deathbed Experiences

Toward the end of the last century the physical sciences and the new theory of evolution were making great progress, but many people felt that science was forcing out the traditional ideas of the spirit and soul. Spiritualism began to flourish, and people flocked to mediums to get in contact with their dead friends and relatives "on the other side." Spiritualists claimed, and indeed still claim, to have found proof of survival.

In 1882, the Society for Psychical Research was founded, and serious research on the phenomena began; but convincing evidence for survival is still lacking over one hundred years later (Blackmore 1988). In 1926, a psychical researcher and Fellow of the Royal Society, Sir William Barrett (1926), published a little book on deathbed visions. The dying apparently saw other worlds before they died and even saw and spoke to the dead. There were cases of music heard at the time of death and reports of attendants actually seeing the spirit leave the body.

With modern medical techniques, deathbed visions like these have become far less common. In those days people died at home with little or no medication and surrounded by their family and friends. Today most people die in the hospital and all too often alone. Paradoxically it is also improved medicine that has led to an increase in quite a different kind of report— that of the near-death experience.

## Close Brushes with Death

Resuscitation from ever more serious heart failure has provided accounts of extraordinary experiences (although this is not the only cause of NDEs). These remained largely ignored until about 15 years ago, when Raymond Moody (1975), an American physician, published his best-selling *Life After Life*. He had talked with many people who had "come back from death," and he put together an account of a typical NDE. In this idealized experience a person hears himself pronounced dead. Then comes a loud buzzing or ringing noise and a long, dark tunnel. He can see his own body from a distance and watch what is happening. Soon he meets others and a "being of light" who shows him a playback of events from his life and helps him to evaluate it. At some point he gets to a barrier and knows that he has to go back. Even though he feels joy, love, and peace there, he returns to his body and life. Later he tries to tell others; but they don't understand, and he soon gives up. Nevertheless the experience deeply affects him, especially his views about life and death.

Many scientists reacted with disbelief. They assumed Moody was at least exaggerating, but he claimed that no one had noticed the experiences before because the patients were too frightened to talk about them. The matter was soon settled by further research. One cardiologist had talked to more than 2,000 people over a period of nearly 20 years and claimed that more than half reported Moody-type experiences (Schoonmaker 1979). In 1982, a Gallup poll found that about 1 in 7 adult Americans had been close to death and about 1 in 20 had had an NDE. It appeared that Moody, at least in outline, was right. In my own research I have come across numerous reports like this one, sent to me by a woman from Cyprus:

An emergency gastrectomy was performed. On the 4th day following that operation I went into shock and became unconscious for several hours. . . Although thought to be unconscious, I remembered, for years afterwards, the entire, detailed conversation that passed between the surgeon and anaesthetist present. . . . I was lying above my own body, totally free of pain, and looking down at my own self with compassion for the agony I could see on the face; I was floating peacefully. Then . . . I was going elsewhere, floating towards a dark, but not frightening, curtain-like area. . . . Then I felt total peace.

Suddenly it all changed—I was slammed back into my body again, very much aware of the agony again.

Within a few years some of the basic questions were being answered. Kenneth Ring (1980), at the University of Connecticut, surveyed 102 people who had come close to death and found almost 50 percent had had what he called a "core experience." He broke this into five stages: peace, body separation, entering the darkness (which is like the tunnel), seeing the light, and entering the light. He found that the later stages were reached by fewer people, which seems to imply that there is an ordered set of experiences waiting to unfold.

One interesting question is whether NDEs are culture specific. What little research there is suggests that in other cultures NDEs have basically the same structure, although religious background seems to influence the way it is interpreted. A few NDEs have even been recorded in children. It is interesting to note that nowadays children are more likely to see living friends than those who have died, presumably because their playmates only rarely die of diseases like scarlet fever or smallpox (Morse et al. 1986).

Perhaps more important is whether you have to be nearly dead to have an NDE. The answer is clearly no (e.g., Morse et al. 1989). Many very similar experiences are recorded of people who have taken certain drugs, were extremely tired, or, occasionally, were just carrying on their ordinary activities.

I must emphasize that these experiences seem completely real—even more real (whatever that may mean) than everyday life. The tunnel experience is not like just imagining going along a tunnel. The view from out of the body seems completely realistic, not like a dream, but as though you really are up there and looking down. Few people experience such profound emotions and insight again during their lifetimes. They do not say, "I've been hallucinating," "I imagined I went to heaven," or "Can I tell you about my lovely dream?" They are more likely to say, "I have been out of my body" or "I saw Grandma in heaven."

Since not everyone who comes close to death has an NDE, it is interesting to ask what sort of people are more likely to have them. Certainly you don't need to be mentally unstable. NDEers do not differ from others in terms of their psychological health or background. Moreover, the NDE does seem to produce profound and positive personality changes (Ring 1984). After this extraordinary experience people claim that they are no longer so motivated by greed and material achievement but are more concerned about other people and their needs. Any theory of the NDE needs to account for this effect.

## Explanations of the NDE

*Astral Projection and the Next World.*
Could we have another body that is the vehicle of consciousness and leaves the physical body at death to go on to another world? This, essentially, is the doctrine of astral projection. In various forms it is very popular and appears in a great deal of New Age and occult literature.

One reason may be that out-of-body experiences (OBEs) are quite common, quite apart from their role in NDEs. Surveys have shown that anywhere from 8 percent (in Iceland) to as much as 50 percent (in special groups, such as marijuana users) have had OBEs at some time during their lives. In my own survey of residents of Bristol I found 12 percent. Typically these people had been resting or lying down and suddenly felt they had left their bodies, usually for no more than a minute or two (Blackmore 1984).

A survey of more than 50 different cultures showed that almost all of them believe in a spirit or soul that could leave the body (Shells 1978). So both the OBE and the belief in another body are common, but what does this mean? Is it just that we cannot bring ourselves to believe that we are nothing more than a mortal body and that death is the end? Or is there really another body?

You might think that such a theory has no place in science and ought to be ignored. I disagree. The only ideas that science can do nothing with are the purely metaphysical ones—ideas that have no measurable consequences and no testable predictions. But if a theory makes predictions, however bizarre, then it can be tested.

The theory of astral projection is, at least in some forms, testable. In the earliest experiments mediums claimed they were able to project their astral bodies to distant rooms and see what was happening. They claimed not to taste bitter aloes on their real tongues, but immediately screwed up their faces in disgust when the substance was placed on their (invisible) astral tongues. Unfortunately these experiments were not properly controlled (Blackmore 1982~.

In other experiments, dying people were weighed to try to detect the astral body as it left. Early this century a weight of about one ounce was claimed, but as the apparatus became more sensitive the weight dropped, implying that it was not a real effect. More recent experiments have used sophisticated detectors of ultraviolet and infrared, magnetic flux or field strength, temperature, or weight to try to capture the astral body of someone having an out-of-body experience. They have even used animals and human "detectors," but no one has yet succeeded in detecting anything reliably (Morris et al. 1978).

If something really leaves the body in OBEs, then you might expect it to be able to see at a distance, in other words to have extrasensory perception (ESP). There have been several experiments with concealed targets. One success was Tart's subject, who lay on a bed with a five-digit number on a shelf above it (Tart 1968). During the night she had an OBE and correctly reported the number, but critics argued that she could have climbed out of the bed to look. Apart from this one, the experiments tend, like so many in parapsychology, to provide equivocal results and no clear signs of any ESP.

So, this theory has been tested but seems to have failed its tests. If there really were astral bodies I would have expected us to have found something out about them by now—other than how hard it is to track them down!

In addition there are major theoretical objections to the idea of astral bodies. If you imagine that the person has gone to another world, perhaps along some "real" tunnel, then you have to ask what relationship there is between this world and the other one. If the other world is an extension of the physical, then it ought to be observable and measurable. The astral body, astral world, and tunnel ought to be detectable in some way, and we ought to be able to say where exactly the tunnel is going. The fact that we can't, leads many people to say the astral world is "on another plane," at a "higher level of vibration," and the like. But unless you can specify just what these mean the ideas are completely empty, even though they may sound appealing. Of course we can never prove that astral bodies don't exist, but my guess is that they probably don't and that this theory is not a useful way to understand OBEs.

*Birth and the NDE.*

Another popular theory makes dying analogous with being born: that the out-of-body experience is literally just that— reliving the moment when you emerged from your mother's body. The tunnel is the birth canal and the white light is the light of the world into which you were born. Even the being of light can be "explained" as an attendant at the birth.

This theory was proposed by Stanislav Grof and Joan Halifax (1977) and popularized by the astronomer Carl Sagan (1979), but it is pitifully inadequate to explain the NDE. For a start the newborn infant would not see anything like a tunnel as it was being born. The birth canal is stretched and compressed and the baby usually forced through it with the top of its head, not with its eyes (which are closed anyway) pointing forward. Also it does not have the mental skills to recognize the people around, and these capacities change so much during growing that adults cannot reconstruct what it was like to be an infant.

"Hypnotic regression to past lives" is another popular claim. In fact much research shows that people who have been hypnotically regressed give the appearance of acting like a baby or a child, but it is no more than acting. For example, they don't make drawings like a real five-year-old would do but like an adult imagines children do. Their vocabulary is too large and in general they overestimate the abilities of children at any given age. There is no evidence (even if the idea made sense) of their "really" going back in time.

Of course the most important question is whether this theory could be tested, and to some extent it can. For example, it predicts that people born by Caesarean section should not have the same tunnel experiences and OBEs. I conducted a survey of people born normally and those born by Caesarean (190 and 36 people, respectively). Almost exactly equal percentages of both groups had had tunnel experiences (36 percent) and OBEs (29 percent). I have not compared the type of birth of people coming close to death, but this would provide further evidence (Blackmore 1982b).

In response to these findings some people have argued that it is not one's own birth that is relived but the idea of birth in general. However, this just reduces the theory to complete vacuousness.

### Just Hallucinations.

Perhaps we should give up and conclude that all the experiences are "just imagination" or "nothing but hallucinations." However, this is the weakest theory of all. The experiences must, in some sense, be hallucinations, but this is not, on its own, any explanation. We have to ask why are they these kinds of hallucinations? Why tunnels?

Some say the tunnel is a symbolic representation of the gateway to another world. But then why always a tunnel and not, say, a gate, doorway, or even the great River Styx? Why the light at the end of the tunnel? And why always above the body, not below it? I have no objection to the theory that the experiences are hallucinations. I only object to the idea that you can explain them by saying, "They are just hallucinations." This explains nothing. A viable theory would answer these questions without dismissing the experiences. That, even if only in tentative form, is what I shall try to provide.

### The Physiology of the Tunnel.

Tunnels do not only occur near death. They are also experienced in epilepsy and migraine, when falling asleep, meditating, or just relaxing, with pressure on both eyeballs, and with certain drugs, such as LSD, psilocybin, and mescaline. I have experienced them many times myself. It is as though the whole world becomes a rushing, roaring tunnel and you are flying along it toward a bright light at the end. No doubt many readers have also been there, for surveys show that about a third of people have—like this terrified man of 28 who had just had the anesthetic for a circumcision.

> I seemed to be hauled at "lightning speed" in a direct line tunnel into
> outer space; (not a floating sensation . . .) but like a rocket at a terrific
> speed. I appeared to have left my body.

In the 1930s, Heinrich Klüver, at the University of Chicago, noted four form constants in hallucinations: the tunnel, the spiral, the lattice or grating, and the cobweb. Their origin probably lies in the structure of the visual cortex, the part of the brain that processes visual information. Imagine that the outside world is mapped onto the back of the eye (on the retina), and then again in the cortex. The mathematics of this mapping (at least to a reasonable approximation) is well known.

Jack Cowan, a neurobiologist at the University of Chicago, has used this mapping to account for the tunnel (Cowan 1982). Brain activity is normally kept stable by some cells inhibiting others. Disinhibition (the reduction of this inhibitory activity) produces too much activity in the brain. This can occur near death (because of lack of oxygen) or with drugs like LSD, which interfere with inhibition. Cowan uses an analogy with fluid mechanics to argue that disinhibition will induce stripes of activity that move across the cortex. Using the mapping it can easily be shown that stripes in the cortex would appear like concentric rings or spirals in the visual world. In other words, if you have stripes in the cortex you will seem to see a tunnel-like pattern of spirals or rings.

This theory is important in showing how the structure of the brain could produce the same hallucination for everyone. However, I was dubious about the idea of these moving stripes, and also Cowan's theory doesn't readily explain the bright light at the center. So Tom Troscianko and I, at the University of Bristol, tried to develop a simpler theory (Blackmore and Troscianko 1989). The most obvious thing about the representation in the cortex is that there are lots of cells representing the center of the visual field but very few for the edges. This means that you can see small things very clearly in the center, but if they are out at the edges you cannot. We took just this simple fact as a starting point and used a computer to simulate what would happen when you have gradually increasing electrical noise in the visual cortex.

The computer program starts with thinly spread dots of light, mapped in the same way as the cortex, with more toward the middle and very few at the edges. Gradually the number of dots increases, mimicking the increasing noise. Now the center begins to look like a white blob and the outer edges gradually get more and more dots. And so it expands until eventually the whole screen is filled with light. The appearance is just like a dark speckly tunnel with a white light at the end, and the light grows bigger and bigger (or nearer and nearer) until it fills the whole screen. (See Figure 1.)

If it seems odd that such a simple picture can give the impression that you are moving, consider two points. First, it is known that random movements in the periphery of the visual field are more likely to be interpreted by the brain as outward than inward movements (Georgeson and Harris 1978). Second, the brain infers our own movement to a great extent from what we see. Therefore, presented with an apparently growing patch of flickering white light your brain will easily interpret it as yourself moving forward into a tunnel.

The theory also makes a prediction about NDEs in the blind. If they are blind because of problems in the eye but have a normal cortex, then they too should see tunnels. But if their blindness stems from a faulty or damaged cortex, they should not. These predictions have yet to be tested.

According to this kind of theory there is, of course, no real tunnel. Nevertheless there is a real physical cause of the tunnel experience. It is noise in the visual cortex. This way we can explain the origin of the tunnel without just dismissing the experiences and without needing to invent other bodies or other worlds.

*Out of the Body Experiences.*

Like tunnels, OBEs are not confined to near death. They too can occur when just relaxing and falling asleep, with meditation, and in epilepsy and migraine. They can also, at least by a few people, be induced at will. I have been interested in OBEs since I had a long and dramatic experience myself (Blackmore 1982a).

It is important to remember that these experiences seem quite real. People don't describe them as dreams or fantasies but as events that actually happened. This is, I presume, why they seek explanations in terms of other bodies or other worlds.

However, we have seen how poorly the astral projection and birth theories cope with OBEs. What we need is a theory that involves no unmeasurable entities or untestable other worlds but explains why the experiences happen; and why they seem so real.

I would start by asking why anything seems real. You might think this is obvious—after all, the things we see out there are real aren't they? Well no, in a sense they aren't. As perceiving creatures all we know is what our senses tell us. And our senses tell us what is "out there" by constructing models of the world with ourselves in it. The whole of the world "out there" and our own bodies are really constructions of our minds. Yet we are sure, all the time, that this construction—if you like, this "model of reality"—is "real" while the other fleeting thoughts we have are unreal. We call the rest of them daydreams, imagination, fantasies, and so on. Our brains have no trouble distinguishing "reality" from "imagination." But this distinction is not given. It is one the brain has to make for itself by deciding which of its own models represents the world "out there." I suggest it does this by comparing all the models it has at any time and choosing the most stable one as "reality."

This will normally work very well. The model created by the senses is the best and most stable the system has. It is obviously "reality," while that image I have of the bar I'm going to go to later is unstable and brief. The choice is easy. By comparison, when you are almost asleep, very frightened, or nearly dying, the model from the senses will be confused and unstable. If you are under terrible stress or suffering oxygen deprivation, then the choice won't be so easy. All the models will be unstable.

So what will happen now? Possibly the tunnel being created by noise in the visual cortex will be the most stable model and so, according to my supposition, this will seem real. Fantasies and imagery might become more stable than the sensory model, and so seem real. The system will have lost input control.

What then should a sensible biological system do to get back to normal? I would suggest that it could try to ask itself—as it were—"Where am I? What is happening?" Even a person under severe stress will have some memory left.

They might recall the accident, or know that they were in hospital for an operation, or remember the pain of the heart attack. So they will try to reconstruct, from what little they can remember, what is happening.

Now we know something very interesting about memory models. Often they are constructed in a bird's-eye view. That is, the events or scenes are seen as though from above. If you find this strange, try to remember the last time you went to a pub or the last time you walked along the seashore. Where are "you" looking from in this recalled scene? If you are looking from above you will see what I mean.

So my explanation of the OBE becomes clear. A memory model in bird's-eye view has taken over from the sensory model. It seems perfectly real because it is the best model the system has got at the time. Indeed, it seems real for just the same reason anything ever seems real.

This theory of the OBE leads to many testable predictions, for example, that people who habitually use bird's-eye views should be more likely to have OBEs. Both Harvey Irwin (1986), an Australian psychologist, and myself (Blackmore 1987) have found that people who dream as though they were spectators have more OBEs, although there seems to be no difference for the waking use of different viewpoints. I have also found that people who can more easily switch viewpoints in their imagination are also more likely to report OBEs.

Of course this theory says that the OBE world is only a memory model. It should only match the real world when the person has already known about something or can deduce it from available information. This presents a big challenge for research on near death. Some researchers claim that people near death can actually see things that they couldn't possibly have known about. For example, the American cardiologist Michael Sabom (1982) claims that patients reported the exact behavior of needles on monitoring apparatus when they had their eyes closed and appeared to be unconscious. Further, he compared these descriptions with those of people imagining they were being resuscitated and found that the real patients gave far more accurate and detailed descriptions.

There are problems with this comparison. Most important, the people really being resuscitated could probably feel some of the manipulations being done on them and hear what was going on. Hearing is the last sense to be lost and, as you will realize if you ever listen to radio plays or news, you can imagine a very clear visual image when you can only hear something. So the dying person could build up a fairly accurate picture this way. Of course hearing doesn't allow you to see the behavior of needles, and so if Sabom is right I am wrong. We can only await further research to find out.

*The Life Review.*

The experience of seeing excerpts from your life flash before you is not really as mysterious as it first seems. It has long been known that stimulation of cells in the temporal lobe of the brain can produce instant experiences that seem like the reliving of memories. Also, temporal-lobe epilepsy can produce similar experiences, and such seizures can involve other limbic structures in the brain, such as the amygdala and hippocampus, which are also associated with memory.

Imagine that the noise in the dying brain stimulates cells like this. The memories will be aroused and, according to my hypothesis, if they are the most stable model the system has at that time they will seem real. For the dying person they may well be more stable than the confused and noisy sensory model.

The link between temporal-lobe epilepsy and the NDE has formed the basis of a thorough neurobiological model of the NDE (Saavedra-Aguilar and Gomez-Jeria 1989). They suggest that the brain stress consequent on the near-death episode leads to the release of neuropeptides and neurotransmitters (in particular the endogenous endorphins). These then stimulate the limbic system and other connected areas. In addition, the effect of the endorphins could account the blissful and other positive emotional states so often associated with the NDE.

Morse provided evidence that some children deprived of oxygen treated with opiates did not have NDE-like hallucinations, and he his colleagues (Morse et al. 1986) have developed a theory based on the role of the neurotransmitter serotonin, rather than the endorphins. Research on the neurochemistry of the NDE is just beginning and should provide us with much more detailed understanding of the life review.

Of course there is more to the review than just memories. The person feels as though she or he is judging these life events, being shown their significance and meaning. But this too, I suggest, is not so very strange. When the normal world of the senses is gone and memories seem real, our perspective on our life changes. We can no longer be attached to our plans, hopes, ambitions, and fears, which fade away and become unimportant, while the past comes to life again. We can only accept it as it is, and there is no one to judge it but ourselves. This is, I think, why so many NDEers say they faced their past life with acceptance and equanimity.

*Other Worlds.*

Now we come to what might seem the most extraordinary parts of the NDE; the worlds beyond the tunnel and OBE. But I think you can now see that they are not so extraordinary at all. In this state the outside world is no longer real, and inner worlds are. Whatever we can imagine clearly enough will seem real. And what will we imagine when we know we are dying? I am sure for many people it is the world they expect or hope to see. Their minds may turn to people they have known who have died before them or to the world they hope to enter next. Like the other images we have been considering, these will seem perfectly real.

Finally, there are those aspects of the NDE that are ineffable—they cannot be put into words. I suspect that this is because some people take yet another step, a step into nonbeing. I shall try to explain this by asking another question. What is consciousness? If you say it is a thing, another body, a substance, you will only get into the kinds of difficulty we got into with OBEs. I prefer to say that consciousness is just what it is like being a mental model. In other words, all the mental models in any person's mind are all conscious, but only one is a model of "me." This is the one that I think of as myself and to which I relate everything else. It gives a core to my life. It allows me to think that I am a person, something that lives on all the time. It allows me to ignore the fact that "I" change from moment to moment and even disappear every night in sleep.

Now when the brain comes close to death, this model of self may simply fall apart. Now there is no self. It is a strange and dramatic experience. For there is no longer an experiencer—yet there is experience.

This state is obviously hard to describe, for the "you" who is trying to describe it cannot imagine not being. Yet this profound experience leaves its mark. The self never seems quite the same again.

*The After Effects.*

I think we can now see why an essentially physiological event can change people's lives so profoundly. The experience has jolted their usual (and erroneous) view of the relationship between themselves and the world. We all too easily assume that we are some kind of persistent entity inhabiting a perishable body. But, as the Buddha taught we have to see through that illusion. The world is only a construction of an information-processing system, and the self is too. I believe that the NDE gives people a glimpse into the nature of their own minds that is hard to get any other way. Drugs can produce it temporarily, mystical experiences can do it for rare people, and long years of practice in meditation or mindfulness can do it. But the NDE can out of the blue strike anyone and show them what they never knew before, that their body is only that— a lump of flesh—that they are not so very important after all. And that is a very freeing and enlightening experience.

*And Afterwards?*

If my analysis of the NDE is correct, we can extrapolate to the next stage. Lack of oxygen first produces increased activity through disinhibition, but eventually it all stops. Since it is this activity that produces the mental models that give rise to consciousness, then all this will cease. There will be no more experience, no more self, and so that, as far as my constructed self is concerned, is the end.

So, are NDEs in or out of the body? I should say neither, for neither experiences nor selves have any location. It is finally death that dissolves the illusion that we are a solid self inside a body.

# AUTHOR'S NOTE

In November 1990 1 visited the Netherlands to give two lectures. The first, on parapsychology, was part of a series organized by the Studium Generale of the

University of Utrecht and titled "Science Confronts the Paranormal." The second was at the Skepsis Conference. Skepsis refers to the very active Dutch skeptics organization called Stichting Skepsis, which means "skeptical foundation." Cornelis de Jager, professor emeritus in astronomy, is the Chair. Skepsis was established in 1987 and publishes the journal *Skepter*. Stichting Skepsis also publishes conference proceedings and monographs on subjects like reincarnation, spiritism, and homeopathy. As its purpose is to educate the public, Skepsis received a starting grant from the government but is now self-supporting, thanks to many generous donations. This is the lecture I presented at the organization's 1990 conference, on "Belief in the Paranormal."

## REFERENCES

**Originally published in *Skeptical Inquirer* 16, no. 1 (1991): 34–45.**

Barrett, W. 1926. Death-bed Visions. London: Methuen.
Blackmore, S. J. 1982a. Beyond the Body. London: Heinemann.
———. 1982b. Birth and the OBE: An unhelpful analogy. Journal of the American Society for Psychical Research, 77:229–238.
———. 1984. A postal survey of OBEs and other experiences. Journal of the Society for Psychical Research, 52:225–244.
———. 1987. Where am I? Perspectives in imagery and the out-of-body experience. Journal of Mental Imagery, 11:53–66.
———. 1988. Do we need a new psychical research? Journal of the Society for Psychical Research, 55:49–59.
Blackmore, S. J., and T. S. Troscianko. 1989. The physiology of the tunnel. Journal of Near-Death Studies, 8:15–28.
Cowan, J. D. 1982. Spontaneous symmetry breaking in large-scale nervous activity International Journal of Quantum Chemistry, 22:1059–1082.
Georgeson, M. A., and M. A. Harris. 1978. Apparent foveo-fugal drift of counterphase gratings. Perception. 7 527–536.
Grof, S., and I Halifax. 1977. The Human Encounter with Death. London: Souvenir Press.
Irwin, H. I. 1986. Perceptual perspectives of visual imagery in OBEs, dreams and reminiscence. Journal of the Society for Psychical Research, 53:210–217.
Moody, R.1975. Life After Life. Covinda, Gal: Mockingbird.
Morris, R. L., S. B. Harary, J. Janis, J. Hartwell, and W. G. Roll. 1978. Studies of communication during out-of-body experiences. Journal of the Society for Psychical Research, 72:1–22.
Morse, J., P. Castillo, D. Venecia, J. Milstein, and D. C. Tyler. 1986. Childhood near-death experiences. American Journal of Diseases of Children, 140:1110–1114.
Morse, I., D. Venecia, and J. Milstein. 1989. Near-death experiences: A neurophysiological explanatory model. Journal of Near-Death Studies, 8 45–53.
Ring, K. 1980. Life at Death. New York: Coward, McCann & Geoghegan.
———. 1986. Heading Toward Omega. New York: Morrow.

Saavedra-Aguilar, J. C., and I S. Gomez-Jeria 1989. Journal of Near-Death Studies, 7: 205–222.

Sabom, M. 1982. Recollections of Death. New York: Harper & Row.

Sagan, C. 1979. Broca's Brain. New York: Random House.

Schoonmaker, F. 1979. Denver cardiologist discloses findings after 18 years of near-death research. Anabiosis, 1:1–2.

Sheils, D. 1978. A cross-cultural study of beliefs in out-of-the-body experiences. Journal of the Society for Psychical Research, 49:697–741.

Tart, C. T. 1978. A psychophysiological study of out-of-the-body experiences in a selected subject. Journal of the Society for Psychical Research, 62:3-27.

# 7

# NUMBER MANIPULATION:
# THE NOT "SO SPOOKY" TIM MCVEIGH

## *Bryan Farha*

Recently I received a mass-distributed e-mail that's been circulating throughout the United States regarding the now-executed convicted mass murderer and terrorist Timothy McVeigh, in which the subject line read, "This is So Spooky!!!" Immediately following is the totality of that mail:

> The total of these dates and times is spooky.
> 04 – The month of the Oklahoma City bombing.
> 19 – The day of the bombing.
> 95 – The bombing year.
> 09 – The hour the bomb went off.
> 02 – The minute the bomb went off.
> 06 – The month McVeigh was executed.
> 11 – The execution day.
> 01 – The execution year.
> 07 – The hour (OK time) he was pronounced dead.
> 14 – The minute (OK time) he was pronounced dead.
> ―――――――――――――――――――――
> Total=168—the number of people killed in the bombing.

At first glance this e-mail might tend to make some people wonder if a supernatural force is trying to send us a message. Eerie? Weird? Spooky? Maybe the ghost of Timothy McVeigh is lurking around us. There is no question that, statistically, this is very coincidental, making it extremely interesting and fun. There are two primary things that make this interesting. First, consistency exists between the two comparison categories—bombing vs. execution—with month, day, year, hour, and minute appearing in each. This is very impressive. Second, of course, is that the total of these "two sets of five" is equal to the number of people killed in the April 19, 1995 bombing of the Alfred P. Murrah Federal building in Oklahoma City at 9:02am—for which McVeigh has since been executed. Or are these numbers truly equal? Are they, in fact, what they appear to be?

## Making the Pool of Numbers Appear Limited

These strategies for number manipulation do not prove that there's not something "spooky" to the ten figures totaling the number of people killed in the bombing, but they may help demonstrate how the number "168" can be "made to fit."

Yes, it's very interesting that those ten numbers totaled 168. Of course, "those" is the key word. It appears as though only ten numbers were looked at and miraculously equaled 168 when added together. But the pool of possible numbers to consider is actually much, much larger than the ones that appear in the e-mail. Hypothetically, had those numbers not totaled 168, perhaps comparing different categories would have made the formula work. Instead of comparing the month, day, year, etc., of the bombing, maybe the number manipulator would have looked at the month, day, year, etc., of the planning of the bombing. Or, numbers could be used surrounding the trial of Timothy McVeigh—using similar factors (month, day, year, etc.) or any one of a number of other factors. Either of these examples would yield entirely different sets of numbers. Further, the figures surrounding convicted co-conspirator Terry Nichols could have been used. The pool of numbers to choose from is practically endless.

## Number Transformation

This involves "fudging" a little bit. Do the numbers presented accurately reflect what is being described, or is some "stretching" taking place? One way numbers or letters can be transformed is to use abbreviations, such as "CA" (California) or "Sept." (September). If true accuracy were to be adhered to, the words "California" or "September" would be used. In the "spooky" e-mail example, what was the true year of the bombing? It was depicted as "95." The actual year of the bombing, of course, was "1995." But it would be pretty difficult to arrive at a figure of "168" by using the accurate figure, so "95" is cleverly chosen. Similarly, the year of the execution is listed as "01," but the actual year was "2001."

## Selective Use of Options

ometimes numbers present themselves in such a way as to give the manipulator a choice. For example, revenue generated can be presented as either "gross" or "net." In the "spooky" e-mail example, notice that the hour (07) McVeigh was pronounced dead is given in Oklahoma time (Central Standard Time). Of course this seems fair since the bombing took place in Oklahoma. But the site for the execution was Terre Haute, Indiana—which is in the Eastern time zone (Eastern Standard Time). To people in Indiana, that number would be "08." Had 169 people been killed, the number manipulator could have still sent the e-mail and made it appear "spooky" simply by using Indiana time.

## Luck

Statistically, strange things will happen in certain situations. Luck helps complete the manipulation puzzle. As many sports experts say, it takes talent, effort, and some luck to win a championship. In the "spooky" e-mail example, notice how enormously different the year of the bombing (95) is compared with the year of the execution (01)—not so different in terms of it being six years later, but in terms of numerical adding of the last two digits when a new century (or millennium) occurs. Because of the new century, a huge figure (95) is added to a tiny figure (01). Had the execution happened before the turn of the century, the number manipulator could never arrive at a figure of "168" after adding all ten figures. Having the opportunity to utilize all four of these number manipulation techniques makes arriving at "168" much easier.

To demonstrate how numbers can be manipulated to appear "spooky," take a look at a few examples this author came up with in just a few hours of number tinkering.

## Arriving at Coinciding Figures
### Considering a Different Number of Casualties

Recalling the "spooky" e-mail, coinciding figures would still be arrived at:

Had 73 been killed—by discarding the bombing year (95).
Had 149 been killed—by discarding the bombing day (19).
Had 154 been killed—by discarding the minute of McVeigh's death (14).
Had 157 been killed—by discarding the execution day (11).
Had 159 been killed—by discarding the bombing month (09).
Had 161 been killed—by discarding the hour of McVeigh's death (07).
Had 162 been killed—by discarding the execution month (06).
Had 164 been killed—by discarding the month of the bombing (04).
Had 166 been killed—by discarding the bombing minute (02).
Had 167 been killed—by discarding the execution year (01).

The above example works for casualties of less than 168. What if the number of people killed were higher? Here's an example of how the clever number manipulator can fix it. Suppose there were 173 casualties. He or she could simply add to 168 the time it took for the first fire truck to arrive on the scene—say five hypothetical minutes. Or a number of other factors could be used, such as the number of minutes it took for the first police car to arrive, or the number of hours it took to uncover survivors, or the actual number of survivors found.

Statistically, the smaller the number, the easier it is to manipulate numbers. So how can larger numbers be manipulated to appear "spooky"? Recall the number transformation employed in the e-mail where the years "95" and "01" were used in place of "1995" and "2001"? Let's see if we can use manipulation while actually employing the larger numbers and have it still appear "spooky." These are actual figures. Try this one:

> 04 – the month of the Oklahoma City bombing.
> 19 – the day of the bombing.
> 1995 – the year of the bombing.
> 09 – the hour the bomb went off.
> 06 – the month McVeigh was executed.
> 2001 – the year McVeigh was executed.
>
> ---
> Total = 4034 – the number of missing FBI document pages related to
>         the case (cbs.com)

Pretty clever, huh? Now let's try one to demonstrate how the pool of numbers is almost endless, and thus we'll use an entirely different set of factors ( *Washington Post* figures):

> 10 – the day McVeigh was indicted by grand jury.
> 02 – the day McVeigh was convicted.
> 06 – the month McVeigh was convicted.
> 1997 – the year McVeigh was convicted.
> 14 – the day McVeigh was formally sentenced.
> 08 – the month McVeigh was sentenced.
> 1997 – the year McVeigh was sentenced.
>
> ---
> Total = 4034 – the number of missing FBI document pages related to
>         the case (cbs.com)

It should now be apparent how easy it is to manipulate numbers. Again, it's possible that something spooky is happening to make those e-mail numbers total 168, but number manipulation is a more plausible explanation. If number manipulation is the answer, it is clearly one of the best this author has ever seen. It's clever, coincidental, fun, entertaining, and many other things. But it's not "spooky."

\*     \*     \*

*MAGIC ELEVEN!!! Pick a number—any number.*

This little gem made the rounds on the Internet after September 11th.

- The date of the attack: 9/11-9+1+1 = 11
- September 11th is the 254th day of the year: 2+5+4 = 11
- After September 11th there are 111 days left until the end of the year.
- 119 is the area code to Iraq/Iran.
- 1+1+9 = 11
- Twin Towers—standing side by side, look like the number 11
- The first plane to hit the towers was Flight 11
- New York State was the 11th State to join the Union
- New York City—11 Letters
- Afghanistan—11 Letters
- The Pentagon—11 Letters
- Ramzi Yousef—11 Letters (convicted of orchestrating the attack on the WTC in 1993)
- Flight 11—92 on board—9+2 = 11
- Flight 77—65 on board—6+5= 11

Oh my God! How worried should I be? I'm going into hiding NOW. See you in a few weeks.

Wait a sec...just realized "YOU CAN'T HIDE" also has 11 letters! What am I gonna do? Help me!!! The terrorists are after me! ME! I can't believe it!

Oh crap, there must be someplace on the planet Earth I could hide! But No..."PLANET EARTH" has 11 letters too!

Maybe Nostradamus can help me. But dare I trust him? There are 11 letters in "NOSTRADAMUS."

I know, the Red Cross can help. No they can't... 11 letters in "THE RED CROSS," ...can't trust them.

I would rely on self defense, but "SELF DEFENSE" has 11 letters in it, too!

Can someone help? Anyone? If so, send me email. No, don't... "SEND ME EMAIL" has 11 letters....

Will this never end? I'm going insane! "GOING INSANE???" Eleven letters!!

Noooooooooooo!!!!!! I guess I'll die alone, even though "I'LL DIE ALONE" has 11 letters....

Oh my God, I just realized that America is doomed! Our Independence Day is July 4th... 7+4=11!

PS: "IT'S BULLSHIT" has 11 letters also.

## NOTE

Originally published in *Skeptic* 9, no. 2 (2002): 12–13.

# 8

# WHY BOGUS THERAPIES OFTEN SEEM TO WORK

## Barry L. Beyerstein

Subtle forces can lead intelligent people (both patients and therapists) to think that a treatment has helped someone when it has not. This is true for new treatments in scientific medicine, as well as for nostrums in folk medicine, fringe practices in "alternative medicine," and the ministrations of faith healers.

Many dubious methods remain on the market primarily because satisfied customers offer testimonials to their worth. Essentially, these people say: "I tried it, and I got better, so it must be effective." The electronic and print media typically portray testimonials as valid evidence. But without proper testing, it is difficult or impossible to determine whether this is so.

There are at least seven reasons why people may erroneously conclude that an ineffective therapy works:

*1. The disease may have run its natural course.* Many diseases are self-limiting. If the condition is not chronic or fatal, the body's own recuperative processes usually restore the sufferer to health. Thus, to demonstrate that a therapy is effective, its proponents must show that the number of patients listed as improved exceeds the number expected to recover without any treatment at all (or that they recover reliably faster than if left untreated). Without detailed records of successes and failures for a large enough number of patients with the same complaint, someone cannot legitimately claim to have exceeded the published norms for unaided recovery.

*2. Many diseases are cyclical.* Such conditions as arthritis, multiple sclerosis, allergies, and gastrointestinal problems normally have "ups and downs." Naturally, sufferers tend to seek therapy during the downturn of any given cycle. In this way, a bogus treatment will have repeated opportunities to coincide with upturns that would have happened anyway.

*3. The placebo effect may be responsible.* Through suggestion, belief, expectancy, cognitive reinterpretation, and diversion of attention, patients given biologically useless treatments often experience measurable relief. Some placebo responses produce actual changes in the physical condition; others are subjective changes that make patients feel better even though there has been no objective change in the underlying pathology.

*4. People who hedge their bets credit the wrong thing.* If improvement occurs after someone has had both "alternative" and science-based treatment, the fringe practice often gets a disproportionate share of the credit.

*5. The original diagnosis or prognosis may have been incorrect.* Scientifically trained physicians are not infallible. A mistaken diagnosis, followed by a trip to a shrine or an "alternative" healer, can lead to a glowing testimonial for curing a condition that would have resolved by itself. In other cases, the diagnosis may be correct but the time frame, which is inherently difficult to predict, might prove inaccurate.

*6. Temporary mood improvement can be confused with cure.* Alternative healers often have forceful, charismatic personalities. To the extent that patients are swept up by the messianic aspects of "alternative medicine," psychological uplift may ensue.

*7. Psychological needs can distort what people perceive and do.* Even when no objective improvement occurs, people with a strong psychological investment in "alternative medicine" can convince themselves they have been helped. According to cognitive dissonance theory, when experiences contradict existing attitudes, feelings, or knowledge, mental distress is produced. People tend to alleviate this discord by reinterpreting (distorting) the offending information. If no relief occurs after committing time, money, and "face" to an alternate course of treatment (and perhaps to the worldview of which it is a part), internal disharmony can result. Rather than admit to themselves or to others that their efforts have been a waste, many people find some redeeming value in the treatment. Core beliefs tend to be vigorously defended by warping perception and memory. Fringe practitioners and their clients are prone to misinterpret cues and remember things as they wish they had happened. They may be selective in what they recall, overestimating their apparent successes while ignoring, downplaying, or explaining away their failures. The scientific method evolved in large part to reduce the impact of this human penchant for jumping to congenial conclusions. In addition, people normally feel obligated to reciprocate when someone does them a good turn. Since most "alternative" therapists sincerely believe they are helping, it is only natural that patients would want to please them in return. Without patients necessarily realizing it, such obligations are sufficient to inflate their perception of how much benefit they have received.

## *Buyer Beware!*

The job of distinguishing real from spurious causal relationships requires well designed studies and logical abstractions from large bodies of data. Many sources of error can mislead people who rely on intuition or informal reasoning to analyze complex events. Before agreeing to any kind of treatment, you should feel confident that it makes sense and has been scientifically validated through studies that control for placebo responses, compliance effects, and judgmental errors. You should be very wary if the "evidence" consists merely of testimonials, self-published pamphlets or books, or items from the popular media.

### NOTE

**Originally published at www.quackwatch.org.**

# 9

# A CLOSE LOOK AT THERAPEUTIC TOUCH

## Linda Rosa, Emily Rosa, Larry Sarner, and Stephen Barrett

THERAPEUTIC TOUCH (TT) is a widely used nursing practice rooted in mysticism but alleged to have a scientific basis. Its practitioners claim to heal or improve many medical problems by manual manipulation of a "human energy field" (HEF) perceptible above the patient's skin. They also claim to detect illnesses and stimulate recuperative powers through their intention to heal. Therapeutic Touch practice guides[1-6] describe 3 basic steps, none of which actually requires touching the patient's body. The first step is centering, in which the practitioner focuses on his or her intent to help the patient. This step resembles meditation and is claimed to benefit the practitioner as well. The second step is assessment, in which the practitioner's hands, from a distance of 5 to 10 cm, sweep over the patient's body from head to feet, "attuning" to the patient's condition by becoming aware of "changes in sensory cues" in the hands. The third step is intervention, in which the practitioner's hands "repattern" the patient's "energy field" by removing "congestion," replenishing depleted areas, and smoothing out ill-flowing areas. The resultant "energy balance" purportedly stems disease and allows the patient's body to heal itself.[7]

Proponents of TT state that they have "seen it work."[8] In a 1995 interview, TT's founder said, "In theory, there should be no limitation on what healing can be accomplished."[9]

## Background

### Professional Recognition

Proponents state that more than 100,000 people worldwide have been trained in TT technique,[38] including at least 43,000 health care professionals,[2] and that about half of those trained actually practice it.[39] Therapeutic Touch is taught in more than 100 colleges and universities in 75 countries.[5] It is said to be the most recognized technique used by practitioners of holistic nursing.[40]

Considered a nursing intervention, it is used by nurses in at least 80 hospitals in North America,[33] often without the permission or even knowledge of attending physicians.[41-43] The policies and procedures books of some institutions recognize TT,[44] and it is the only treatment for the "energy-field disturbance" diagnosis is recognized by the North American Nursing Diagnosis Association.[45] *RN,* one of the nursing profession's largest periodicals, has published many articles favorable to TT.[46-52]

Many professional nursing organizations promote TT. In 1987, the 50,000-member Order of Nurses of Quebec endorsed TT as a "bona fide" nursing skill.[32] The National League for Nursing, the credentialing agency for nursing schools in the United States, denies having an official stand on TT but has promoted it through books and videotapes,[3, 53, 54] and the league's executive director and a recent president are prominent advocates.[55] The American Nurses' Association holds TT workshops at its national conventions. Its official journal published the premier articles on TT[56-59] as well as a recent article designated for continuing education credits.[60] The association's immediate past president has written editorials defending TT against criticism.[61] The American Holistic Nursing Association offers certification in "healing touch," a TT variant.[62] The Nurse Healers and Professional Associates Cooperative, which was formed to promote TT, claims about 1200 members.[39]

### The TT Hypothesis

Therapeutic Touch was conceived in the early 1970s by Dolores Krieger, PhD, RN, a faculty member at New York University's Division of Nursing. Although often presented as a scientific adaptation of "laying-on of hands,"[63-68] TT is imbued with metaphysical ideas.

Krieger initially identified TT's active agent as prana, an ayurvedic, or traditional Indian, concept of "life force." She stated,

"Health is considered a harmonious relationship between the individual and his total environment. There is postulated a continuing interacting flow of energies from within the individual outward, and from the environment to the various levels of the individual. Healing, it is said, helps to restore this equilibrium in the ill person. Disease, within this context, is considered an indication of a disturbance in the free flow of the pranic current."

Krieger further postulated that this "pranic current" can be controlled by the will of the healer.

"When an individual who is healthy touches an ill person with the intent of helping or healing him, he acts as a transference agent for the flow of prana from himself to the ill person. It was this added input of prana . . . that helped the ill person to overcome his illness or to feel better, more vital."

Others associate all this with the Chinese notion of *qi,* a "life energy" alleged to flow through the human body through invisible "meridians." Those inspired by mystical healers of India describe this energy as flowing in and out of sites of the body that they call chakras.

Soon after its conception, TT became linked with the westernized notions of the late Martha Rogers, dean of nursing at New York University. She asserted that humans do not merely possess energy fields but are energy fields and constantly interact with the "environmental field" around them. Rogers dubbed her approach the "Science of Unitary Man,"[69] which later became known as the more neutral "Science of Unitary Human Beings." Her nomenclature stimulated the pursuit of TT as a "scientific" practice. Almost all TT discussion today is based on Rogers' concepts, although Eastern metaphysical terms such as chakra[2, 70] and yin-yang[71] are still used.

The HEF postulated by TT theorists resembles the "magnetic fluid" or "animal magnetism" postulated during the 18th century by Anton Mesmer and his followers. Mesmerism held that illnesses are caused by obstacles to the free flow of this fluid and that skilled healers ("sensitives") could remove these obstacles by making passes with their hands. Some aspects of mesmerism were revived in the 19th century by Theosophy, an occult religion that incorporated Eastern metaphysical concepts and underlies many current New Age ideas.[72] Dora Kunz, who is considered TT's codeveloper, was president of the Theosophical Society of America from 1975 to 1987. She collaborated with Krieger on the early TT studies and claims to be a fifth-generation "sensitive" and a "gifted healer."[20]

Therapeutic Touch is set apart from many other alternative healing modalities, as well as from scientific medicine, by its emphasis on the healer's intention. Whereas the testing of most therapies requires controlling for the placebo effect (often influenced by the recipient's belief about efficacy), TT theorists suggest that the placebo effect is irrelevant. According to Krieger,

"Faith on the part of the subject does not make a significant difference in the healing effect. Rather, the role of faith seems to be psychological, affecting his acceptance of his illness or consequent recovery and what this means to him. The healer, on the other hand, must have some belief system that underlies his actions, if one is to attribute rationality to his behavior."[65]

Thus, the TT hypothesis and the entire practice of TT rest on the idea that the patient's energy field can be detected and intentionally manipulated by the therapist. With this in mind, early practitioners concluded that physical contact might not be necessary.[13] The thesis that the HEF extends beyond the skin and can be influenced from several centimeters away from the body's surface is said to have been tested by Janet Quinn, PhD, and reported in her 1982 dissertation.[14] However, that study merely showed no difference between groups of patients who did or did not have actual contact during TT. Although Quinn's work has never been substantiated, nearly all TT practitioners today use only the noncontact form of TT.

As originally developed by Krieger, TT did involve touch, although clothes and other materials interposed between practitioner and patient were not considered significant.[56] It was named TT because the aboriginal term laying-on of hands was considered an obstacle to acceptance by "curriculum committees and other institutional bulwarks of today's society."[66] The mysticism has been downplayed, and various scientific-sounding mechanisms have been proposed. These include the therapeutic value of skin-to-skin contact, electron transfer resonance, oxygen uptake by hemoglobin, stereochemical similarities of hemoglobin and chlorophyll, electrostatic potentials influenced by healer brain activity, and unspecified concepts from quantum theory.[66, 67]

Therapeutic Touch is said to be in the vanguard of treatments that allow "healing" to take place, as opposed to the "curing" pejoratively ascribed to mainstream medical practice. Therapeutic Touch supposedly requires little training beyond refining an innate ability to focus one's intent to heal; the patient's body then does the rest.[5] Nurses who claim a unique professional emphasis on caring are said to be specially situated to help patients by using TT.[56, 59] Nonetheless, proponents also state that nearly everyone has an innate ability to learn TT, even small children and juvenile delinquents on parole.[2, 17, 32]

Proponents describe the HEF as real and perceptible. Reporting on a pilot study, Krieger claimed that 4 blindfolded men with transected spinal cords "could tell exactly where the nurse's hands were in their HEFs during the Therapeutic Touch interaction."[5] In ordinary TT sessions, practitioners go through motions that supposedly interact with the patient's energy field, including flicking "excess energy" from their fingertips.[3]

Therapeutic Touch is claimed to have only beneficial effects.[39] However, some proponents warn against overly lengthy sessions or overtreating certain areas of the body. This caution is based on the notion that too much energy can be imparted to a patient, especially an infant, which could lead to hyperactivity.[5,73,74]

## Methods

In 1996 and 1997, by searching for advertisements and following other leads, 2 of us (L.R. and L.S.) located 25 TT practitioners in northeastern Colorado, 21 of whom readily agreed to be tested. Of those who did not, 1 stated she was not qualified, 2 gave no reason, and 1 agreed but canceled on the day of the test.

The reported practice experience of those tested ranged from 1 to 27 years. There were 9 nurses, 7 certified massage therapists, 2 laypersons, 1 chiropractor, 1 medical assistant, and 1 phlebotomist. All but 2 were women, which reflects the sex ratio of the practitioner population. One nurse had published an article on TT in a journal for nurse practitioners.

There were 2 series of tests. In 1996, 15 practitioners were tested at their homes or offices on different days for a period of several months. In 1997, 13 practitioners, including 7 from the first series, were tested in a single day.

The test procedures were explained by 1 of the authors (E.R.), who designed the experiment herself. The first series of tests was conducted when she was 9 years old. The participants were informed that the study would be published as her fourth-grade science-fair project and gave their consent to be tested. The decision to submit the results to a scientific journal was made several months later, after people who heard about the results encouraged publication. The second test series was done at the request of a Public Broadcasting Service television producer who had heard about the first study. Participants in the second series were informed that the test would be videotaped for possible broadcast and gave their consent.

During each test, the practitioners rested their hands, palms up, on a flat surface, approximately 25 to 30 cm apart. To prevent the experimenter's hands from being seen, a tall, opaque screen with cutouts at its base was placed over the subject's arms, and a cloth towel was attached to the screen and draped over them.

To examine whether air movement or body heat might be detectable by the experimental subjects, preliminary tests were performed on 7 other subjects who had no training or belief in TT. Four were children who were unaware of the purpose of the test. Those results indicated that the apparatus prevented tactile cues from reaching the subject.

The odds of getting 8 of 10 trials correct by chance alone is 45 of 1024 ($P=.04$), a level considered significant in many clinical trials. We decided in advance that an individual would "pass" by making 8 or more correct selections and that those passing the test would be retested, although the retest results would not be included in the group analysis. Results for the group as a whole would not be considered positive unless the average score was above 6.7 at a 90% confidence level.

## Results

### Initial Test Results

If HEF perception through TT was possible, the experimental subjects should have each been able to detect the experimenter's hand in 10 (100%) of 10 trials. Chance alone would produce an average score of 5 (50%).

Before testing, all participants said they could use TT to significant therapeutic advantage. Each described sensory cues they used to assess and manipulate the HEF. All participants but 1 certified massage therapist expressed high confidence in their TT abilities, and even the aforementioned certified massage therapist said afterward that she felt she had passed the test to her own satisfaction.

In the initial trial, the subjects stated the correct location of the investigator's hand in 70 (47%) of 150 tries. The number of correct choices ranged from 2 to 8. Only 1 subject scored 8, and that same subject scored only 6 on the retest.

After each set of trials, the results were discussed with the participant. Because all but 1 of the trials could have been considered a failure, the participants usually chose to discuss possible explanations for failure. Their rationalizations included the following: (1) The experimenter left a "memory" of her hand behind, making it increasingly difficult in successive trials to detect the real hand from the memory. However, the first attempts (7 correct and 8 incorrect) scored no better than the rest. Moreover, practitioners should be able to tell whether a field they are sensing is "fresh." (2) The left hand is the "receiver" of energy and the right hand is the "transmitter." Therefore, it can be more difficult to detect the field when it is above the right hand. Of the 72 tests in which the hand was placed above the subjects' right hand, only 27 (38%) had correct responses. In addition, 35 (44%) of 80 incorrect answers involved the allegedly more receptive left hand-consistent with randomness. Moreover, practitioners customarily use both hands to assess. (3) Subjects should be permitted to identify the experimenter's field before beginning actual trials. Each subject could be given an example of the experimenter hovering her hand above each of theirs and told which hand it is. Since the effects of the HEF are described in unsubtle terms, such a procedure should not be necessary, but including it would remove a possible post hoc objection. Therefore, we did so in the follow-up testing. (4) The experimenter should be more proactive, centering herself and/or attempting to transmit energy through her own intentionality. This contradicts the fundamental premise of TT, since the experimenter's role is analogous to that of a patient. Only the practitioner's intentionality and preparation (centering) are theoretically necessary. If not so, the early experiments (on relatively uninvolved subjects, such as infants and barley seeds), cited frequently by TT advocates, must also be discounted. (5) Some subjects complained that their hands became so hot after a few trials that they were no longer able to sense the experimenter's HEF or they experienced difficulty doing so. This explanation clashes with TT's basic premise that practitioners can sense and manipulate the HEF with their hands during sessions that typically last 20 to 30 minutes. If practitioners become insensitive after only brief testing, the TT hypothesis is untestable. Those who made this complaint did so after they knew the results, not before. Moreover, only 7 of the 15 first trials produced correct responses.

*Follow-up Test Results*

The 1997 testing was completed in 1 day and videotaped by a professional film crew. Each subject was allowed to "feel" the investigator's energy field and choose which hand the investigator would use for testing. Seven subjects chose her left hand, and 6 chose her right hand.

The test results were similar to those of the first series. The subjects correctly located the investigator's hand in only 53 (41%) of 130 tries. The number of correct answers ranged from 1 to 7. After learning of their test scores, one participant said he was distracted by the towel over his hands, another said that her hands had been too dry, and several complained that the presence of the television crew had made it difficult to concentrate and/or added to the stress of the test. However, we do not believe that the situation was more stressful or distracting than the settings in which many hospital nurses practice TT (eg, intensive care units).

## Comment

Practitioners of TT are generally reluctant to be tested by people who are not proponents. In 1996, the James Randi Educational Foundation offered $742 000 to anyone who could demonstrate an ability to detect an HEF under conditions similar to those of our study. Although more than 40 000 American practitioners claim to have such an ability, only 1 person attempted the demonstration. She failed, and the offer, now more than $1.1 million, has had no further volunteers despite extensive recruiting efforts.[129]

We suspect that the present authors were able to secure the cooperation of 21 practitioners because the person conducting the test was a child who displayed no skepticism.

## Conclusion

Therapeutic touch is grounded on the concept that people have an energy field that is readily detectable (and modifiable) by TT practitioners. However, this study found that 21 experienced practitioners, when blinded, were unable to tell which of their hands was in the experimenter's energy field. The mean correct score for the 28 sets of 10 tests was 4.4, which is close to what would be expected for random guessing.

To our knowledge, no other objective, quantitative study involving more than a few TT practitioners has been published, and no well-designed study demonstrates any health benefit from TT. These facts, together with our experimental findings, suggest that TT claims are groundless and that further use of TT by health professionals is unjustified.

## REFERENCES

Originally published in *Journal of the American Medical Association* 279, no. 13 (April 1998): 1005-1010. Copyright 1998, American Medical Association. All rights reserved.

1. Boguslawski M. The use of Therapeutic Touch in nursing. *J Continuing Educ Nurs.* 1979;10(4):9-15.

2. Krieger D. *Therapeutic Touch Inner Workbook.* Santa Fe, NM: Bear; 1997:162.

3. Quinn JF. *Therapeutic Touch: Healing Through Human Energy Fields: Theory and Research* [videotapes and study guide]. New York, NY: National League for Nursing; 1994:42-2485-42-2487, 42-2493.

4. Krieger D. *Living the Therapeutic Touch: Healing as a Lifestyle.* New York, NY: Dodd Mead; 1987.

5. Krieger D. *Accepting Your Power to Heal: The Personal Practice of Therapeutic Touch.* Santa Fe, NM: Bear; 1993.

6. Chiappone J. *The Light Touch: An Easy Guide to Hands-on Healing.* Lake Mary, Fla: Holistic Reflections; 1989:14.

7. Quinn JF, Strelkauskas AJ. Psycho immunologic effects of Therapeutic Touch on practitioners and recently bereaved recipients: a pilot study. *ANS Adv Nurs Sci.* 1993;15(4):13-26.

8. Jarboux D. Nurse knows Therapeutic Touch "works." *Boulder Sunday Camera.* January 2, 1994:3E.

9. Putnam ZE. Using consciousness to heal. *Massage Ther J.* Fall 1995:47-48, 50, 52, 54, 56, 58, 60.

10. Leduc E. Therapeutic Touch. *Neonat Network.* 1987;5(6):46-47. Library Holdings Bibliographic Links

11. Krieger D. Therapeutic Touch during childbirth preparation by the Lamaze method and its relation to marital satisfaction and state anxiety of the married couple. In: Krieger D. *Living the Therapeutic Touch: Healing as a Lifestyle.* New York, NY: Dodd Mead; 1987:157-187.

12. Glazer S. The mystery of "Therapeutic Touch." *Washington Post.* December 19-26, 1995; Health section:16-17.

13. Heidt PR. Effect of Therapeutic Touch on anxiety level of hospitalized patients. *Nurs Res.* 1981;30(1):32-37.

14. Quinn JF. *An Investigation of the Effects of Therapeutic Touch Done Without Physical Contact on State Anxiety of Hospitalized Cardiovascular Patients* [dissertation]. New York: New York University; 1982.

15. Thayer MB. Touching with intent: using Therapeutic Touch. *Pediatr Nurs.* 1990;16(1):70-72. Library Holdings Bibliographic Links

16. Mersmann CA. *Therapeutic Touch and Milk Let Down in Mothers of Non-nursing Preterm Infants* [dissertation]. New York: New York University; 1993.

17. France NEM. The child's perception of the human energy field using Therapeutic Touch. *J Holistic Nurs.* 1993;11:319-331.

18. Meehan MTC. The Science of Unitary Human Beings and theory-based practice: Therapeutic Touch. In: Barrett EAM, ed. Visions *of Rogers' Science-Based Nursing.* New York, NY: National League for Nursing; 1990:67-81. Publication 15-2285.

19. Peters PJ. The *Lifestyle Changes of Selected Therapeutic Touch Practitioners: An Oral History* [dissertation]. Minneapolis, Minn: Walden University; 1992.

20. Boguslawski M. Therapeutic Touch: a facilitator of pain relief. *Top Clin Nurs.* 1980;2(1):27-37.

21. Satir F. Healing hands. *Olympian.* July 19, 1994.

22. Brown PR. *The Effects of Therapeutic Touch on Chemotherapy-induced Nausea and Vomiting: A Pilot Study* [master's thesis]. Reno: University of Nevada; 1981.

23. Sodergren KA. *The Effect of Absorption and Social Closeness on Responses to Educational and Relaxation Therapies in Patients With Anticipatory Nausea and Vomiting During Cancer Chemotherapy* [dissertation]. Minneapolis: University of Minnesota; 1993.

24. Dollar CE. *Effects of Therapeutic Touch on Perception of Pain and Physiological Measurements From Tension Headache in Adults: A Pilot Study* [master's thesis]. Jackson: University of Mississippi Medical Center; 1993.

25. Quinn JF. Holding sacred space: the nurse as healing environment. *Holistic Nurs Pract.* 1992;6(4):26-36. Library Holdings Bibliographic Links

26. Wirth DP. The effect of non-contact Therapeutic Touch on the healing rate of full thickness dermal wounds. *Subtle Energies.* 1990;1(1):1-20.

27. Woods DL. *The Effect of Therapeutic Touch on Disruptive Behaviors of Individuals With Dementia of the Alzheimer Type* [master's thesis]. Seattle: University of Washington; 1993.

28. Misra MM. *The Effects of Therapeutic Touch on Menstruation.* [master's thesis]. Long Beach: California State University; 1993.

29. Putnam ZE. The woman behind Therapeutic Touch: Dolores Krieger, PhD, RN. *Massage Ther J.* Fall 1995:50, 52.

30. Simington JA, Laing GP. Effects of Therapeutic Touch on anxiety in the institutionalized elderly. *Clin Nurs Res.* 1993;2:438-450.

31. Quinn JF. The Senior's Therapeutic Touch Education Program. *Holistic Nurs Pract.* 1992;7(1):32-37. Library Holdings Bibliographic Links

32. Krieger D. Therapeutic Touch: two decades of research, teaching and clinical practice. *Imprint.* 1990;37:83, 86-88.

33. Fiely D. Field of beams. *Columbus Dispatch.* August 20, 1995:1B-2B.

34. Calvert R. Dolores Krieger, PhD, and her Therapeutic Touch. *Massage Magazine.* 1994;47:56-60.

35. Mueller-Jackson ME. The use of Therapeutic Touch in the nursing care of the terminally ill person. In: Borelli MD, Heidt PR, eds. *Therapeutic Touch: A Book of Readings.* New York, NY: Springer; 1981:72-79.

36. Brunjes CAF. Therapeutic Touch: a healing modality throughout life. *Top Clin Nurs.* 1983;5(2):72-79. Library Holdings Bibliographic Links

37. Messenger TC, Roberts KT. The terminally ill: serenity nursing interventions for hospice clients. *J Gerontol Nurs.* 1994;20(11):17-22.

38. Maxwell J. Nursing's new age? *Christianity Today.* 1996;40(3):96-99.

39. Kauffold MP. TT: healing or hokum?. debate over "energy medicine" runs hot. *Chicago Tribune Nursing News.* November 19, 1995:1.

40. Keegan L. Holistic nursing. *J Post Anesth Nurs.* 1989;4(1):17-21.

41. Bullough VL, Bullough B. Therapeutic Touch: why do nurses believe? *Skeptical Inquirer.* 1993;17:169-174.

42. Dr Quinn studies Therapeutic Touch. *University of Colorado School of Nursing News* May 1989:1.

43. Cabico LL. *A Phenomenological Study of the Experiences of Nurses Practicing Therapeutic Touch* [master's thesis]. Buffalo, NY: D'Youville College; 1992.

44. Rosa LA. When magic gets to play science. *Rocky Mountain Skeptic.* 1993;10(6):10-12.

45. Carpenito LJ. *Nursing Diagnosis: Application to Clinical Practice.* 6th ed. Philadelphia, Pa: Lippincott; 1995:355-358.

46. Sandroff R. A skeptic's guide to Therapeutic Touch. *RN.* 1980;43(1):24-30, 82-83.

47. Raucheisen ML. Therapeutic Touch: maybe there's something to it after all. *RN.* 1984;47(12):49-51.

48. Haddad A. Acute care decisions: ethics in action. *RN.* 1994;57(11):21-22, 24.

49. Swackhamer AH. It's time to broaden our practice. *RN.* 1995;58(1):49-51.

50. Schmidt CM. The basics of Therapeutic Touch. *RN.* 1995;58(6):50, 52, 54.

51. Ledwith SP. Therapeutic Touch and mastectomy: a case study. *RN.* 1995;58(7):51-53.

52. Keegan L, Cerrato PL. Nurses are embracing holistic healing. *RN.* 1996;59(4):59.

53. Moccia P. *New Approaches to Theory Development.* New York, NY: National League for Nursing; 1986;15-1992.

54. Barrett EAM. *Visions of Rogers' Science-Based Nursing.* New York, NY: National League for Nursing; 1990;15-2285.

55. Moccia P. Letter to the editor. *Time.* 1994;144(24):18.

56. Krieger D. Therapeutic Touch: the imprimatur of nursing. *Am J Nurs.* 1975;75:784-787.

57. Krieger D, Peper E, Ancoli S. Therapeutic Touch: searching for evidence of physiological change. *Am J Nurs.* 1979;79:660-662.

58. Macrae JA. Therapeutic Touch in practice. *Am J Nurs.* 1979;79:664-665.

59. Quinn JF. One nurse's evolution as a healer. *Am J Nurs.* 1979;79:662-664.

60. Mackey RB. Discover the healing power of Therapeutic Touch. *Am J Nurs.* 1995;95(4):27-33.

61. Joel LA. Alternative solutions to health problems. *Am J Nurs.* 1995;95(7):7.

62. Hover-Kramer D. Healing Touch certificate program continues to bring the human dimension to the nation's nurses. *Beginnings.* 1992;12(2):3.

63. Cowens C, Monte T. *A Gift for Healing: How You Can Use Therapeutic Touch.* New York, NY: Crown Publishing Group; 1996.

64. Krieger D. The response of in-vivo human hemoglobin to an active healing therapy by direct laying on of hands. *Human Dimensions.* Autumn 1972:12-15.

65. Krieger D. Therapeutic Touch and healing energies from the laying on of hands. *J Holistic Health.* 1975;1:23-30.

66. Krieger D. Therapeutic Touch: an ancient, but unorthodox nursing intervention. *J N Y State Nurs Assoc.* 1975;6(2):6-10.

67. Krieger D. Healing by the laying-on of hands as a facilitator of bio-energetic change: the response of in-vivo human hemoglobin. *Int J Psychoenergy Syst.* 1976;1(1):121-129.

68. Krieger D. The relationship of touch, with intent to help or to heal, to subjects' in-vivo hemoglobin values: a study in personalized interaction. In: *Proceedings of the Ninth ANA Nurses Research Conference.* New York, NY: American Nurses' Association; 1973:39-59.

69. Rogers ME. *An Introduction to the Theoretical Basis of Nursing.* Philadelphia, Pa: Davis; 1970.

70. Karagulla S, Kunz D. *The Chakras and the Human Energy Field: Correlations Between Medical Science & Clairvoyant Observation.* Wheaton, Ill: Theosophical Publishing House; 1989.

71. Randolph GL. The yin and yang of clinical practice. *Top Clin Nurs.* 1979;1(1):31-42.

72. Brierton TD. Employers' New Age training programs fail to alter the consciousness of the EEOC. *Lab Law J.* 1992;43:411-420.

73. Emery CE. Therapeutic Touch: healing technique or New Age rite?. *Providence Sun Journal-Bulletin.* November 27, 1994:A1, A24.

74. Knaster M. Dolores Krieger's Therapeutic Touch. *East/West*. 1989;19(8):54-57, 59, 79-80.

75. Colorado State Board of Nursing. *Subcommittee to Investigate the Awarding of Continuing Education Units to Nurses for the Study of Therapeutic Touch and Other Non-traditional and Complementary Healing Modalities. Recommendations*. Denver: Colorado State Board of Nursing; 1992.

76. Mulloney SS, Wells-Federman C. Therapeutic Touch: a healing modality. *J Cardiovascul Nurs*. 1996;10(3):27-49.

77. Brown CC, Fischer R, Wagman AMI, Horrom N, Marks P. The EEG in meditation and Therapeutic Touch healing. *J Altered States Conscious*. 1977;3:169-180.

78. Quinn JF. Therapeutic Touch as energy exchange: testing the theory. *Adv Nurs Sci*. 1984;6(1):42-49.

79. Guerrero MA. *The Effects of Therapeutic Touch on State-Trait Anxiety Level of Oncology Patients* [master's thesis]. Galveston: University of Texas; 1985.

80. Keller EAK, Bzdek VM. Effects of Therapeutic Touch on tension headache pain. *Nurs Res*. 1986;35(2):101-106.

81. Meehan MTC. Theory development. In: Barrett EAM, ed. *Visions of Rogers' Science-Based Nursing*. New York, NY: National League for Nursing; 1990;15-2285:197-207.

82. Kramer NA. Comparison of Therapeutic Touch and Casual Touch in stress reduction of hospitalized children. *Pediatr Nurs*. 1990;16:483-485.

83. Shuzman E. *The Effect of Trait Anxiety and Patient Expectation of Therapeutic Touch on the Reduction in State Anxiety in Preoperative Patients Who Receive Therapeutic Touch* [dissertation]. New York: New York University; 1993.

84. Sies MM. *An Exploratory Study of Relaxation Response in Nurses Who Utilize Therapeutic Touch* [master's thesis]. East Lansing: Michigan State University; 1993.

85. Wirth DP, Richardson JT, Eidelman WS, O'Malley AC. Full thickness dermal wounds treated with Therapeutic Touch: a replication and extension. *Complementary Ther Med*. 1993;1:127-132.

86. Gagne D, Toye RC. The effects of Therapeutic Touch and relaxation therapy in reducing anxiety. *Arch Psych Nurs*. 1994;8:184-189.

87. Schlotfeldt RM. Critique of: Krieger D. The relationship of touch, with intent to help or heal, to subjects' in-vivo hemoglobin values: a study in personalized interaction. In: *Proceedings of the Ninth ANA Nurses Research Conference*. New York, NY: American Nurses' Association; 1973:59-65.

88. Walike BC, Bruno P, Donaldson S, et al. ". . . [A]ttempts to embellish a totally unscientific process with the aura of science. . . ." *Am J Nurs*. 1975;75:1275, 1278, 1292..

89. Levine ME. "The science is spurious. . . ." *Am J Nurs*. 1979;79:1379-1380.

90. Clark PE, Clark MJ. Therapeutic Touch: is there a scientific basis for the practice?. *Nurs Res*. 1984;33(1):38-41.

91. Meehan MTC. Therapeutic Touch. In: Bulechek GM, McCloskey JC, eds. *Nursing Interventions: Essential Nursing Treatments*. 2nd ed. Philadelphia, Pa: Saunders; 1992:201-212.

92. Fish S. Therapeutic Touch: can we trust the data?. *J Christian Nurs*. 1993;10(3):6-8.

93. Meehan MTC. Therapeutic Touch and postoperative pain: a Rogerian research study. *Nurs Sci Q* 1993;6(2):69-78.

94. Bandman EL, Bandman B. *Critical Thinking in Nursing*. Norwalk, Conn: Appleton & Lange; 1995.

95. Meehan MTC. Quackery and pseudo-science. *Am J Nurs.* 1995;75(7):17.

96. Meehan MTC. . . . And still more on TT. *Res Nurs Health.* 1995;18:471-472.

97. Rosa LA. *Survey of Therapeutic Touch "Research.".* Loveland, Colo: Front Range Skeptics; 1996.

98. Tharnstrom CAL. *The Effects of Non-contact Therapeutic Touch on the Parasympathetic Nervous System as Evidenced by Superficial Skin Temperature and Perceived Stress* [master's thesis]. San Jose, Calif: San Jose State University; 1993.

99. Parkes BS. *Therapeutic Touch as an Intervention to Reduce Anxiety in Elderly Hospitalized Patients* [dissertation]. Austin: University of Texas; 1985.

100. Mueller Hinze ML. *The Effects of Therapeutic Touch and Acupressure on Experimentally-Induced Pain* [dissertation]. Austin: University of Texas; 1988.

101. Bowers DP. *The Effects of Therapeutic Touch on State Anxiety and Physiological Measurements in Preoperative Clients* [master's thesis]. San Jose, Calif: San Jose State University, 1992.

102. Olson M, Sneed NV. Anxiety and Therapeutic Touch. *Issues Ment Health Nurs.* 1995;16:97-108.

103. Fedoruk RB. *Transfer of the Relaxation Response: Therapeutic Touch as a Method for Reduction of Stress in Premature Neonates* [dissertation]. Baltimore: University of Maryland; 1984.

104. Hogg PK. *The Effects of Acupressure on the Psychological and Physiological Rehabilitation of the Stroke Patient* [dissertation]. Alameda: California School of Professional Psychology; 1985.

105. Snyder M, Egan EC, Burns KR. Interventions for decreasing agitation behaviors in persons with dementia. *J Gerontol Nurs.* 1995;21(7):34-40, 54-55.

106. Schweitzer SF. *The Effects of Therapeutic Touch on Short-term Memory Recall in the Aging Population: A Pilot Study* [master's thesis]. Reno: University of Nevada; 1980.

107. Randolph GL. Therapeutic and physical touch: physiological response to stressful stimuli. *Nurs Res.* 1984;33(1):33-36.

108. Nodine JL. *The Effects of Therapeutic Touch on Anxiety and Well-being in Third Trimester Pregnant Women* [master's thesis]. Tucson: University of Arizona; 1987.

109. Post NW. *The Effects of Therapeutic Touch on Muscle Tone* [master's thesis]. San Jose, Calif: San Jose State University; 1990.

110. Straneva JAE. *Therapeutic Touch and In Vitro Erythropoiesis* [dissertation]. Bloomington: Indiana University; 1992.

111. Bush AM, Geist CR. Testing electromagnetic explanations for a possible psychokinetic effect of Therapeutic Touch in germinating corn seed. *Psycholog Rep.* 1992;70:891-896.

112. Edge H. The effect of laying on of hands on an enzyme: an attempted replication. Paper presented at: 22nd Annual Convention of the Parapsychology Association; August 15-18, 1979; Moraga, Calif.

113. Meehan MTC. *The Effect of Therapeutic Touch on the Experience of Acute Pain in Postoperative Patients* [dissertation]. New York: New York University; 1985.

114. Hale EH. *A Study of the Relationship Between Therapeutic Touch and the Anxiety Levels of Hospitalized Adults* [dissertation]. Denton: Texas Women's University; 1986.

115. Quinn JF. Therapeutic Touch as energy exchange: replication and extension. *Nurs Sci Q.* 1989;2(2):79-87.

116. Claman HN, Freeman R, Quissel D, et al. *Report of the Chancellor's Committee on Therapeutic Touch.* Denver: University of Colorado Health Sciences Center; 1994.

117. Butgereit B. Therapeutic Touch: UAB to study controversial treatment for Pentagon. *Birmingham News.* November 17, 1994:1A, 10A.

118. Turner JG. *Tri-Service Nursing Research Grant Proposal* [revised abstract]. 1994. Grant No. MDA905-94-Z-0080.

119. Turner JG. *The Effect of Therapeutic Touch on Pain & Anxiety in Burn Patients* [grant final report]. Tri-Service Nursing Research Program; November 14, 1996. Grant No. N94020.

120. Lionberger HJ. Therapeutic Touch: a healing modality or a caring strategy. In: Chinn PL, ed. *Nursing Research Methodology: Issues and Implementation.* Rockville, Md: Aspen Publishers; 1986:169-180.

121. Lionberger HJ. *An Interpretive Study of Nurses' Practice of Therapeutic Touch* [dissertation]. San Francisco: University of California; 1985.

122. Polk SH. *Client's Perceptions of Experiences Following the Intervention Modality of Therapeutic Touch* [master's thesis]. Tempe: Arizona State University; 1985.

123. Hamilton-Wyatt GK. *Therapeutic Touch: Promoting and Assessing Conceptual Change Among Health Care Professionals* [dissertation]. East Lansing: Michigan State University; 1988.

124. Heidt PR. Openness: a qualitative analysis of nurses' and patients' experiences of Therapeutic Touch. *Image J Nurs Sch.* 1990;22:180-186.

125. Thomas-Beckett JG. *Attitudes Toward Therapeutic Touch: A Pilot Study of Women With Breast Cancer* [master's thesis]. East Lansing: Michigan State University; 1991.

126. Clark AJ, Seifert P. Client perceptions of Therapeutic Touch. Paper presented at: Third Annual West Alabama Conference on Clinical Nursing Research; 1992.

127. Samarel N. The experience of receiving Therapeutic Touch. *J Adv Nurs.* 1992;17:651-657.

128. Hughes PP. *The Experience of Therapeutic Touch as a Treatment Modality With Adolescent Psychiatric Patients* [master's thesis]. Albuquerque: University of New Mexico; 1994.

129. James Randi Educational Foundation. Available at: http://www.randi.org Accessed March 15, 1997.

# 10

## CHIROPRACTIC'S DIRTY SECRET: NECK MANIPULATION AND STROKES

### Stephen Barrett

Stroke from chiropractic neck manipulation occurs when an artery to the brain ruptures or becomes blocked as a result of being stretched. The injury often results from extreme rotation in which the practitioner's hands are placed on the patient's head in order to rotate the cervical spine by rotating the head.[1] The vertebral artery is vulnerable because it winds around the topmost cervical vertebra (atlas) to enter the skull, so that any abrupt rotation may stretch the artery and tear its delicate lining. The anatomical problem is illustrated on page 7 of *The Chiropractic Report*, July 1999. A blood clot formed over the injured area may subsequently be dislodged and block a smaller artery that supplies the brain. Less frequently, the vessel may be blocked by blood that collects in the vessel wall at the site of the dissection.[2]

Chiropractors would like you to believe that the incidence of stroke following neck manipulation is extremely small. Speculations exist that the odds of a serious complication due to neck manipulation are somewhere between one in 40,000 and one in 10 million manipulations. No one really knows, however, because (a) there has been little systematic study of its frequency; (b) the largest malpractice insurers won't reveal how many cases they know about; and (c) a large majority of cases that medical doctors see are not reported in scientific journals.

### Published Reports

In 1992, researchers at the Stanford Stroke Center asked 486 California members of the American Academy of Neurology how many patients they had seen during the previous two years who had suffered a stroke within 24 hours of neck manipulation by a chiropractor. The survey was sponsored by the American Heart Association. A total of 177 neurologists reported treating 56 such patients, all of whom were between the ages of 21 and 60. One patient had died, and 48 were left with permanent neurologic deficits such as slurred speech, inability to arrange words properly, and vertigo (dizziness). The usual cause of the strokes was thought to be a tear between the inner and outer walls of the vertebral arteries, which caused the arterial walls to balloon and block the flow of blood to the brain. Three of the strokes involved tears of the carotid arteries.[3] In 1991, according to circulation figures from *Dynamic Chiropractic*, California had about 19% of the chiropractors practicing in the United States, which suggests that about 147 cases of stroke each year were seen by neurologists nationwide. Of course, additional cases could have been seen by other doctors who did not respond to the survey.

A 1993 review concluded that potential complications and unknown benefits indicate that children should not undergo neck manipulation.[4]

Louis Sportelli, DC, NCMIC president and a former ACA board chairman contends that chiropractic neck manipulation is quite safe. In an 1994 interview reported by the Associated Press, he reacted to the American Heart Association study by saying, "I yawned at it. It's old news." He also said that other studies suggest that chiropractic neck manipulation results in a stroke somewhere between one in a million and one in three million cases.[5] The one-in-a-million figure could be correct if California's chiropractors had been averaging about 60 neck manipulations per week. Later that year, during a televised interview with "Inside Edition," Sportelli said the "worst-case scenario" was one in 500,000 but added: "When you weigh the procedure against any other procedure in the health-care industry, it is probably the lowest risk factor of anything." According to the program's narrator, Sportelli said that 90% of his patients receive neck manipulation.

In 1996, RAND issued a booklet that tabulated more than 100 published case reports and estimated that the number of strokes, cord compressions, fractures, and large blood clots was 1.46 per million neck manipulations. Even though this number appears small, it is significant because many of the manipulations chiropractors do should not be done. In addition, as the report itself noted, neither the number of manipulations performed nor the number of complications has been systematically studied.[6] Since some people are more susceptible than others, it has also been argued that the incidence should be expressed as rate per patient rather than rate per adjustment.

In 1996, the National Chiropractic Mutual Insurance Company (NCMIC), which is the largest American chiropractic malpractice insurer, published a report called "Vertebrobasilar Stroke Following Manipulation," written by Allen G.J. Terrett, an Australian chiropractic educator/researcher. Terrett based his findings on 183 cases of vertebrobasilar strokes (VBS) reported between 1934 and 1994. He concluded that 105 of the manipulations had been administered by a chiropractor, 25 were done by a medical practitioner, 31 had been done by another type of practitioner, and that the practitioner type for the remaining 22 was not specified in the report. He concluded that VBS is "very rare," that current pretesting procedures are seldom able to predict susceptibility, and that in 25 cases serious injury might have been avoided if the practitioner had recognized that symptoms occurring after a manipulation indicated that further manipulations should not be done.[7]

A 1999 review of 116 articles published between 1925 and 1997 found 177 cases of neck injury associated with neck manipulation, at least 60% of which was done by chiropractors.[8]

In 2001, NCMIC published a second edition of Terrett's book, titled, "Current Concepts: Vertebrobasilar Complications following Spinal Manipulation," which covered 255 cases published between 1934 and 1999.[9] NCMIC's Web site claims that the book "includes an analysis of every known case related to this subject." That description is not true. It does not include many strokes that resulted in lawsuits against NCMIC policyholders but were not published in scientific journals. And it does not include the thoroughly documented case of Kristi Bedenbauer, whose autopsy report I personally mailed to Terrett after speaking with him in 1995.

In 2001, Canadian researchers published a report about the relationships between chiropractic care and the incidence of vertebrovascular accidents (VBAs) due to vertebral artery dissection or blockage in Ontario, Canada, between 1993 and 1998. Using hospital records, each of 582 VBA cases was age- and sex-matched to four controls with no history of stroke. Health insurance billing records were used to document use of chiropractic services. The study found that VBA patients under age 45 were five times more likely than controls to (a) have visited a chiropractor within a week of the VBA and (b) to have had three or more visits with neck manipulations. No relationship was found after age 45. The authors discuss possible shortcomings of the study and urge that further research be done.[10] An accompanying editorial states that the data correspond to an incidence of 1.3 cases of vertebral artery dissection or blockage per 100,000 individuals receiving chiropractic neck manipulation, a number higher than most chiropractic estimates.[11]

In 2001, British researchers reported on a survey in which all members of the Association of British Neurologists were asked to report cases referred to them of neurological complications occurring within 24 hours of neck manipulation over a 12-month period. The 35 reported cases included 7 strokes involving the vertebrobasilar artery and 2 strokes involving a carotid artery. None of the 35 cases were reported to medical journals.[12] Edzard Ernst, professor of complementary medicine at the University of Exeter School of Sport and Health Sciences, believes that these results are very significant. In a recent commentary, he stated:

> One gets the impression that the risks of spinal manipulation are being played down, particularly by chiropractors. Perhaps the best indication that this is true are estimates of incidence rates based on assumptions, which are unproven at best and unrealistic at worse. One such assumption, for instance, is that 10% of actual complications will be reported. Our recent survey, however,

demonstrated an underreporting rate of 100%. This extreme level of underreporting obviously renders estimates nonsensical.[13]

In 2002, researchers representing the Canadian Stroke Consortium reported on 98 cases in which external trauma ranging from "trivial" to "severe" was identified as the trigger of strokes caused by blood clots formed in arteries supplying the brain. Chiropractic-style neck manipulation was the apparent cause of 38 of the cases, 30 involving vertebral artery dissection and 8 involving carotid artery dissection. Other Canadian statistics indicate the incidence of ischemic strokes in people under 45 is about 750 a year. The researchers believe that their data indicate that 20% are due to neck manipulation, so there may be "gross underreporting" of chiropractic manipulation as a cause of stroke.[14]

In 2003, another research team reviewed the records of 151 patients under age 60 with cervical arterial dissection and ischemic stroke or transient ischemic attack (TIA) from between 1995 and 2000 at two academic stroke centers. After an interview and a blinded chart review, 51 patients with dissection and 100 control patients were studied. Patients with dissection were more likely to have undergone spinal manipulation within 30 days (14% vs 3%). The authors concluded that spinal manipulation is associated with vertebral arterial dissection and that a significant increase in neck pain following neck manipulation warrants immediate medical evaluation.[15]

In 2006, the *Journal of Neurology* published a German Vertebral Artery Dissection Study Group report about 36 patients who had experienced vertebral artery dissection associated with neck manipulation.[16] Twenty-six patients developed their symptoms within 48 hours after a manipulation, including five patients who got symptoms at the time of manipulation and four who developed them within the next hour. In 27 patients, special imaging procedures confirmed that blood supply had decreased in the areas supplied by the vertebral arteries as suggested by the neurological examinations. In all but one of the 36 patients, the symptoms had not previously occurred and were clearly distinguishable from the complaints that led them to seek manipulative care. This report is highly significant but needs careful interpretation. Although it is titled "Vertebral dissections after chiropractic neck manipulation "only four of the patients were actually manipulated by chiropractors. Half were treated by orthopedic surgeons, five by a physiotherapist, and the rest by a neurologist, general medical practitioner, or homeopath. It is possible—although unlikely—that the nonchiropractors used techniques that were more dangerous than chiropractors use in North America. The authors suggested that the orthopedists' treatment was safer, but there is no way to determine this from their data. Regardless, the study supports the assertion that neck manipulation can cause strokes—which many chiropractors deny.

## *Are Complications Predictable?*

Although some chiropractors advocate "screening tests" with the hope of detecting individuals prone to stroke due to neck manipulation.[17,18] These tests, which include holding the head and neck in positions of rotation to see whether the patient gets dizzy, are not reliable[19], partly because manipulation can rotate the neck further than can be done with the tests.[19] Listening over the neck arteries with a stethoscope to detect a murmur, for example, has not been proven reliable, though patients that have one should be referred to a physician. Vascular function tests in which the patient's head is briefly held in the positions used during cervical manipulation are also not reliable as a screen for high-risk patients because a thrust that further stretches the vertebral artery could still damage the vessel wall." In a chapter in the leading chiropractic textbook, Terrett and a colleague have stated:

> Even after performing the relevant case history, physical examination, and vertebrobasilar function tests, accidents may still occur. There is no conclusive, foolproof screening procedure to eliminate patients at risk. Most victims are young, without [bony] or vascular pathology, and do not present with vertebrobasilar symptoms. The screening procedures described cannot detect those patients in whom [manipulation] may cause an injury. They give a false sense of security to the practitioner.[20]

Several medical reports have described chiropractic patients who, after neck manipulation, complained of dizziness and other symptoms of transient loss of blood supply to the brain but were manipulated again and had a full-blown stroke. During a workshop I attended at the 1995 Chiropractic Centennial Celebration, Terrett said such symptoms are ominous and that chiropractors should abandon rotational manipulations that overstretch the vertebral arteries. But, as far as I know, his remarks have not been published and have had no impact on his professional colleagues.

The lack of predictability has been supported by data published by Scott Haldeman, DC, MD, PhD, a chiropractor who has served as an expert witness (usually for the defense) in many court cases involving chiropractic injury. In 1995, he published an abstract summarizing his review of 53 cases that had not been previously reported in medical or chiropractic journals. His report stated:

> These cases represent approximately a 45% increase in the number of such cases reported in the English language literature over the past 100 years. . . . No clear cut risk factors can be elicited from the data. Previously proposed risk factors such as migraine headaches, hypertension, diabetes, history of cardiovascular disease, oral contraceptives, recent head or neck trauma, or abnormalities on x-rays do not appear to be significantly greater in patients who have

cerebrovascular complications of manipulation than that noted in the general population.[21]

Haldeman's main point was he could not identify any factor that could predict that a particular patient was prone to cerebrovascular injury from neck manipulation. This report was published in the proceedings of 1995 Chiropractic Centennial Celebration and was not cited in either the RAND or NCMIC reports.

In 2001, Haldeman and two colleagues published a more detailed analysis that covered 64 cases involving malpractice claims filed between 1978 and 1994 [22]. They reported that 59 (92%) came to treatment with a history of head or neck symptoms. However, the report provides insufficient information to judge whether manipulation could have been useful for treating their condition. Of course, malpractice claims don't present the full story, because most victims of professional negligence do not take legal action. Even when serious injury results, some are simply not inclined toward suing, some don't blame the practitioner, some have an aversion to lawyers, and some can't find an attorney willing to represent them.

## What Should Be Done?

Chiropractors cannot agree among themselves whether the problem is significant enough to inform patients that vertebrobasilar stroke is a possible complication of manipulation.[19,23] In 1993, the Canadian Chiropractic Association published a consent form which stated, in part:

> Doctors of chiropractic, medical doctors, and physical therapists using manual therapy treatments for patients with neck problems such as yours are required to explain that there have been rare cases of injury to a vertebral artery as a result of treatment. Such an injury has been known to cause stroke, sometimes with serious neurological injury. The chances of this happening are extremely remote, approximately 1 per 1 million treatments.
>
> Appropriate tests will be performed on you to help identify if you may be susceptible to that kind of injury. . . .[24]

This notice is a step in the right direction but does not go far enough. A proper consent should disclose that (a) the risk is unknown; (b) alternative treatments may be available; (c) in many cases, neck symptoms will go away without treatment; (d) certain types of neck manipulation carry a higher risk than others; and (e) claims that spinal manipulation can remedy systemic diseases, boost immunity, improve general health, or prolong life have neither scientific justification nor a plausible rationale.

In 2003, a coroner's jury concluded that Lana Dale Lewis of Toronto, Canada, was killed in 1996 by a chiropractic neck manipulation. Among other things, the jury recommended that all patients for whom neck manipulation is recommended be informed that risk exists and that the Ontarion Ministry of Health establish a database for chiropractors and other health professionals to report on neck adjustments.[25]

## The Bottom Line

As far as I know, most chiropractors do not warn their patients that neck manipulation entails risks. I believe they should and that the profession should implement a reporting system that would enable this matter to be appropriately studied. This might be achieved if (a) state licensing boards required that all such cases be reported, and (b) chiropractic malpractice insurance companies, which now keep their data secret, were required to disclose them to an independently operated database that has input from both medical doctors and chiropractors.

Meanwhile, since stroke is such a devastating event, every effort should be made to stop chiropractors from manipulating necks without adequate reason. Many believe that all types of headaches might be amenable to spinal manipulation even though no scientific evidence supports such a belief. Many include neck manipulation as part of "preventative maintenance" that involves unnecessarily treating people who have no symptoms. Even worse, some chiropractors—often referred to as "upper cervical specialists"—claim that most human ailments are the result of misalignment of the topmost vertebrae (atlas and axis) and that every patient they see needs neck manipulation. Neck manipulation of children under age 12 should be outlawed.[26]

## REFERENCES

Originally published at www.quackwatch.org.

1. Homola S. *Inside Chiropractic: A Patient's Guide*. Amherst, NY: Prometheus Books, 1999.
2. Norris JW and others. "Sudden neck movement and cervical artery dissection." *Canadian Medical Journal* 163:38–40, 2000.
3. Lee KP and others. "Neurologic complications following chiropractic manipulation: A survey of California neurologists." *Neurology* 45:1213–1215, 1995.
4. Powell FC and others. "A risk/benefit analysis of spinal manipulation therapy for relief of lumbar or cervical pain." *Neurosurgery* 33:73–79, 1993.
5. Haney DQ. "Twist of the neck can cause stroke warn doctors." Associated Press news release, Feb 19, 1994.

6.    Coulter I and others. *The Appropriateness of Manipulation and Mobilization of the Cervical Spine.* Santa Monica, CA: RAND, 1996, pp. 18–43.

7.    Di Fabio R. "Manipulation of the cervical spine: Risks and benefits." *Physical Therapy* 79:50–65, 1999.

8.    Terrett AGJ. *Current Concepts in Vertebrobasilar Stroke following Manipulation.* West Des Moines, IA: National Chiropractic Mutual Insurance Company, Inc., 2001.

9.    Terrett AGJ. *Current Concepts: Vertebrobasilar Complications following Spinal Manipulation.* West Des Moines, IA: NCMIC Group, Inc., 2001.

10.   Rotherwell DAM and others. "Chiropractic manipulation and stroke." *Stroke* 32:1054–1059, 2001.

11.   Bousser MG. Editorial comment. *Stroke* 32:1059–1060, 2001.

12.   Stevinson C and others. "Neurological complications of cervical spine manipulation." *Journal of the Royal Society of Medicine* 94:107–110, 2001.

13.   Ernst E. "Spinal manipulation: Its safety is uncertain." *Canadian Medical Association Journal* 166:40–41, 2002.

14.   Beletsky V. "Chiropractic manipulation may be underestimated as cause of stroke." Presented at the American Stroke Association's 27th International Stroke Conference, San Antonio, Texas, February 7–8, 2002.

15.   Smith WS and others. "Spinal manipulative therapy is an independent risk factor for vertebral artery dissection." *Neurology* 60:1424–1428, 2003.

16.   Reuter U and others. "Vertebral artery dissections after chiropractic neck manipulation in Germany over three years." *Journal of Neurology* 256:724–730, 2006.

17.   George PE and others. "Identification of high-risk pre-stroke patient." *ACA Journal of Chiropractic* 15:S26–S28, 1981.

18.   Sullivan EC. "Prevent strokes: Screening can help." *The Chiropractic Journal,* May 1989, 27.

19.   Chapman-Smith D. "Cervical adjustment: Rotation is fine, pre-testing is out, but get consent." *The Chiropractic Report* 13(4):1–3, 6–7, 1999.

20.   Terrett AGJ, Kleynhans AM. "Cerebrovascular complications of manipulation." In Haldeman S (ed). *Principles and Practice of Chiropractic, Second Edition.* East Norwalk, CT: Appleton and Lange, 1992.

21.   Haldeman S, Kohlbeck F, McGregor M. "Cerebrovascular complications following cervical spine manipulation therapy: A review of 53 cases," Conference Proceedings of the Chiropractic Centennial, July 6–8, 1995, 282–283. Davenport, IA: Chiropractic Centennial Foundation, 1995.

22.   Haldeman S and others. "Unpredictability of cerebrovascular ischemia associated with cervical spine manipulation therapy." *Spine* 27:49–55, 2001.

23.   Magner G. "Informed consent is needed." In Magner G. *Chiropractic: The Victim's Perspective.* Amherst, NY: Prometheus Books, 1995, 177–184.

24.   Henderson D et al. "Clinical Guidelines for Chiropractic Practice in Canada." Toronto: Canadian Chiropractic Association, 1994, 4.

25.   "Coroner's jury concludes that neck manipulation killed Canadian woman." *Chirobase,* Jan 22, 2004.

26.   Stewart B and others. "Statement of concern to the Canadian public from Canadian neurologists regarding the debilitating and fatal damage manipulation of the neck may cause to the nervous system." February 2002.

# 11

## MAGNET THERAPY: A SKEPTICAL VIEW

### *Stephen Barrett*

During the past several years, magnetic devices have been claimed to relieve pain and to have therapeutic value against a large number of diseases and conditions. The way to evaluate such claims is to ask whether scientific studies have been published. Pulsed electromagnetic fields—which induce measurable electric fields—have been demonstrated effective for treating slow-healing fractures and have shown promise for a few other conditions. However, few studies have been published on the effect on pain of small, static magnets marketed to consumers.[1] Explanations that magnetic fields "increase circulation," "reduce inflammation," or "speed recovery from injuries" are simplistic and are not supported by the weight of experimental evidence.[2]

The main basis for the claims is a double-blind test study, conducted at Baylor College of Medicine in Houston, which compared the effects of magnets and sham magnets on knee pain. The study involved 50 adult patients with pain related to having been infected with the polio virus when they were children. A static magnetic device or a placebo device was applied to the patient's skin for 45 minutes. The patients were asked to rate how much pain they experienced when a "trigger point was touched." The researchers reported that the 29 patients exposed to the magnetic device achieved lower pain scores than did the 21 who were exposed to the placebo device.[3] Although this study is cited by nearly everyone selling magnets, it provides no legitimate basis for concluding that magnets offer any health-related benefit:

- Although the groups were said to be selected randomly, the ratio of women to men in the experimental group was twice that of the control group. If women happen to be more responsive to placebos than men, a surplus of women in the "treatment" group would tend to improve that group's score.

- The age of the placebo group was four years higher than that of the control group. If advanced age makes a person more difficult to treat, the "treatment" group would again have a scoring advantage.
- The investigators did not measure the exact pressure exerted by the blunt object at the trigger point before and after the study.
- Even if the above considerations have no significance, the study should not be extrapolated to suggest that other types of pain can be relieved by magnets.
- There was just one brief exposure and no systematic follow-up of patients. Thus there was no way to tell whether any improvement would be more than temporary.
- The authors themselves acknowledge that the study was a "pilot study." Pilot studies are done to determine whether it makes sense to invest in a larger more definitive study. They never provide a legitimate basis for marketing any product as effective against any symptom or health problem.

Two better-designed, longer-lasting pain studies have been negative:

- Researchers at the New York College of Podiatric Medicine have reported negative results in a study of patients with heel pain. Over a 4-week period, 19 patients wore a molded insole containing a magnetic foil, while 15 patients wore the same type of insole with no magnetic foil. In both groups, 60% reported improvement, which suggests that the magnetic foil conveyed no benefit.[4]
- More recently, researchers at the VA Medical Center in Prescott, Arizona conducted a randomized, double-blind, placebo-controlled, crossover study involving 20 patients with chronic back pain. Each patient was exposed to real and sham bipolar permanent magnets during alternate weeks, for 6 hours per day, 3 days per week for a week, with a 1-week period between the treatment weeks. No difference in pain or mobility was found between the treatment and sham-treatment periods.[5]

Magnets have also been claimed to increase circulation. This claim is false. If it were true, placing a magnet on the skin would make the area under the magnet become red, which it does not. Moreover, a well-designed study that actually measured blood flow has found no increase. The study involved 12 healthy volunteers who were exposed to either a 1000-gauss magnetic disk or an identically appearing disk that was not magnetic. No change in the amount or speed of blood flow was observed when either disk was applied to their arm.[6] The magnets were manufactured by Magnetherapy, Inc, of Riviera Beach, Florida, a company that has been subjected to two regulatory actions.

## Legal and Regulatory Actions

In 1998, Magnetherapy, Inc., signed an Assurance of Voluntary Compliance with the State of Texas to pay a $30,000 penalty and to stop claiming that wearing its magnetic device near areas of pain and inflammation will relieve pain due to arthritis, migraine headaches, sciatica or heel spurs. The agreement also requires Magnetherapy to stop making claims that its magnets can cure, treat, or mitigate any disease or can affect any change in the human body, unless its devices are FDA-approved for those purposes.[7] Ads for the company's Tectonic Magnets had featured testimonials from athletes, including golfers from the senior pro tours. Various ads had claimed that Tectonic Magnets would provide symptomatic relief from certain painful conditions and could restore range of motion to muscles and joints. The company had provided retailers with display packages that included health claims, written testimonials, and posters of sports stars. Texas Attorney General Dan Morales stated that some claims were false or unsubstantiated and others had rendered the product unapproved medical devices under Texas law. In 1997, the FDA had warned Magnetherapy to stop claiming that its products would relieve arthritis; tennis elbow; low back pain; sciatica; migraine headache; muscle soreness; neck, knee, ankle, and shoulder pain; heel spurs; bunions; arthritic fingers and toes; and could reduce pain and inflammation in the affected areas by increasing blood and oxygen flow.[8]

In 1999, the FTC obtained a consent agreement barring two companies from making unsubstantiated claims about their magnetic products. Magnetic Therapeutic Technologies, of Irving, Texas, is barred from claiming that its magnetic sleep pads or other products: (a) are effective against cancers, diabetic ulcers, arthritis, degenerative joint conditions, or high blood pressure; (b) could stabilize or increase the T-cell count of HIV patients; (c) could reduce muscle spasms in persons with multiple sclerosis; (d) could reduce nerve spasms associated with diabetic neuropathy; (e) could increase bone density, immunity, or circulation; or (f) are comparable or superior to prescription pain medicine. Pain Stops Here! Inc., of Baiting Hollow, N.Y., may no longer claim that its "magnetized water" or other products are useful against cancer, diseases of the liver or other internal organs, gallstones, kidney stones, urinary infection, gastric ulcers, dysentery, diarrhea, skin ulcers, bed sores, arthritis, bursitis, tendinitis, sprains, strains, sciatica, heart disease, circulatory disease, arthritis, auto-immune illness, neuro-degenerative disease, and allergies, and could stimulate the growth of plants.

On August 8, 2000, the Consumer Justice Center, of Laguna Niguel, California filed suit in Orange County Superior Court charging that Florsheim and a local shoe store (Shoe Emporium) made false and fraudulent claims that their MagneForce shoes (a) correct "magnetic deficiency," (b) "generate a deep-penetrating magnetic field which increases blood circulation; reduces leg and back fatigue; and provides natural pain relief and improved energy level."; and

(c) their claims are established and proven by scientific studies.[9] A few days after this suit was filed, Florsheim removed the disputed ad from its Web site.

In 2001, Richard Markoll, his wife Ernestine, David H. Trock, M.D., and Bio-Magnetic Treatment Systems (BMTS) pled guilty to criminal charges in connection with a scheme involving pulsed magnetic therapy. The participants used fraudulent billing codes to seek payment from Medicare and three other insurance plans for treatment with a device (Electro-Magnetic Induction Treatment System, Model 30/30) that lacked FDA approval.[10] The treatments— called pulsed signal therapy (PST)—were administered in a clinical trial on an investigational basis not approved by the FDA. The Markolls were sentenced to 3 years probation, a $4,000 fine and a $100 special assessment. Ernestine Markoll was sentenced to 2 years probation, a $1,000 fine and a $25 special assessment. Magnetic Therapy, was sentenced to a 1-day summary probation and a $200 special assessment. The Markolls also signed a civil settlement under which they agreed to pay the U.S Government $4 million.[11] The device was invented by Richard Markoll, MD, PhD, who does not have a medical license but is described in Web site biographies as a graduate of Grace University School of Medicine, a Caribbean medical school. Trock, a former principal investigator for Magnetic Therapy Center, PC, Danbury, CT, was sentenced to 6 months probation. and ordered to make restitution of $35,250.[12] Trock has co-authored studies claiming that PST is effective for treating pain, but the device is not FDA-approved for that purpose.

In September 2002, California Attorney General Bill Lockyer charged Florida-based European Health Concepts, Inc. (EHC) with making false and misleading claims about its magnetic mattress pads and seat cushions. The complaint, filed in Sacramento Superior Court, also named EHC president Kevin Todd and several sales managers and agents as defendants. The suit seeks more than $1 million in civil penalties for engaging in unfair business practices and making false claims; $500,000 in civil penalties for transactions involving senior citizens; and full restitution for purchasers of the products. The complaint alleged that prospective customers, primarily senior citizens, were invited to attend a free dinner seminar at which they were told that EHC's products could help people suffering from fibromyalgia, lupus, sciatica, herniated discs, asthma, bronchitis, cataracts, chronic fatigue syndrome, colitis, diverticulitis, heart disease, multiple sclerosis, and more than 50 other health conditions. The sales agents offered phony price discounts for immediate purchases that actually were the company's regular prices.[13]

## The Bottom Line

There is no scientific basis to conclude that small, static magnets can relieve pain or influence the course of any disease. In fact, many of today's products produce no significant magnetic field at or beneath the skin's surface.

# REFERENCES

**Originally published at www.quackwatch.org.**

1. Livingston JD. "Magnetic therapy: Plausible attraction." *Skeptical Inquirer* 25-30, 58, 1998.

2. Ramey DW. "Magnetic and electromagnetic therapy." *Scientific Review of Alternative Medicine* 2(1):13-19, 1998.

3. Vallbona C, Hazelwood CF, Jurida G. "Response of pain to static magnetic fields in postpolio patients: A double-blind pilot study." *Archives of Physical and Rehabilitative Medicine* 78:1200-1203, 1997.

4. Caselli MA and others. "Evaluation of magnetic foil and PPT Insoles in the treatment of heel pain." *Journal of the American Podiatric Medical Association* 87:11-16, 1997.

5. Collacott EA and others. "Bipolar permanent magnets for the treatment of chronic low back pain." *JAMA* 283:1322-1325, 2000.

6. Mayrovitz HN and others. "Assessment of the short-term effects of a permanent magnet on normal skin blood circulation via laser-Doppler flowmetry." *Scientific Review of Alternative Medicine* 6(1):9-12, 2002.

7. Morales halts unproven claims for magnet therapy. News release, April 9, 1998.

8. Gill LJ. Letter to William L. Roper, Feb 3, 1997.

9. *Jeff Wynton and the Consumer Justice Center v. Florsheim Group, Inc., Shoe Emporium.* Superior Court of California, Orange County, Case #00CC09419, filed Aug 8, 2000.

10. Burns EB. "Omnibus ruling on defendants' motion to strike and motions to dismiss." *United States of America v Richard Markoll*, Ernestine Binder Markoll, and Bio-Magnetic Systems, Inc. U.S. District Court, District of Connecticut, No. 3:00cr133(EBB), Jan 2001.

11. Defense Criminal Investigative Service press release, Aug, 2001.

12. Defense Criminal Investigative Service press release, June, 2001.

13. Barrett S. "California Attorney General sues magnetic mattress pad sellers." *Quackwatch*, Sept 24, 2002.

# 12

# THE GREAT DILUTION DELUSION

## *James Randi*

On November 27, 2002, the BBC-TV program *Horizon*, featuring a test of homeopathy, was broadcast in the UK. Many viewers expressed their conviction that we'd heard the death-knell of this form of quackery; I disagreed. To explain my reluctance to join the funeral procession, I offer readers this:

Oliver Wendell Holmes (1809-1894) was a celebrated physician, poet, humorist and professor of anatomy and physiology at Harvard, as well as the father of O.W.H. Junior (1841-1935), who became a renowned justice of the U.S. Supreme Court. In 1842, Senior wrote an essay, "Homeopathy and Its Kindred Delusions," which had originally been presented by him as two lectures to the Boston Society for the Diffusion of Useful Knowledge. I present here two excerpts from the essay, to illustrate just how little the situation has changed in the last 160 years.

> In 1835 a public challenge was offered to the best-known Homeopathic physician in Paris to select any ten substances asserted to produce the most striking effects; to prepare them himself; to choose one by lot without knowing which of them he had taken, and try it upon himself or an intelligent and devoted Homeopathist, and, waiting his own time, to come forward and tell what substance had been employed. The challenge was at first accepted, but the acceptance was retracted before the time of trial arrived.

Sound familiar? In April of 1999, Nobel laureate Brian Josephson publicly challenged the American Physical Society (APS) to conduct tests of the claims of Dr. Jacques Benveniste in regard to homeopathy, at the same time predicting that the APS would fear to do so. I advised the APS to accept Josephson's challenge, and they did so. They also offered to pay all costs of the tests. From that day to this—three years and seven months ago—we have not heard from either Brian Josephson, or Jacques Benveniste....

Holmes Senior concluded:

> From all this I think it fair to conclude that the catalogues of symptoms attributed in Homeopathic works to the influence of various drugs upon healthy persons are not entitled to any confidence.

Exactly the decision I came to, long ago. But read on. Holmes, in his essay, described the thorough manner in which homeopathic claims had been examined. He compared the eventual results to those met with when magical "tractors"—devices said to "withdraw" diseases and invented by Dr. Elisha Perkins in 1801—were clearly shown to be pure quackery and yet persisted in Holmes' time. Holmes:

> Now to suppose that any trial can absolutely silence people, would be to forget the whole experience of the past. Dr. Haygarth and Dr. Alderson could not stop the sale of the five-guinea Tractors, although they proved that they could work the same miracles with pieces of wood and tobacco-pipe. It takes time for truth to operate, as well as Homoeopathic globules. Many persons thought the results of these trials were decisive enough of the nullity of the treatment; those who wish to see the kind of special pleading and evasion by which it is attempted to cover results which, stated by the Homoeopathic Examiner itself, look exceedingly like a miserable failure, may consult the opening flourish of that Journal. I had not the intention to speak of these public trials at all, having abundant other evidence on the point. But I think it best, on the whole, to mention two of them in a few words—the one instituted at Naples and that of Andral.

There have been few names in the medical profession, for the last half century, so widely known throughout the world of science as that of M. Esquirol, whose life was devoted to the treatment of insanity, and who was without a rival in that department of practical medicine. It is from an analysis communicated by him to the Gazette Médicale de Paris that I derive my acquaintance with the account of the trial at Naples by Dr. Panvini, physician to the Hospital della Pace. This account seems to be entirely deserving of credit. Ten patients were set apart, and not allowed to take any [homeopathic] medicine at all,—much against the wish of the Homoeopathic physician.

All of them got well, and of course all of them would have been claimed as triumphs if they had been submitted to the treatment. Six other slight cases (each of which is specified) got well under the Homoeopathic treatment—but with none of its asserted specific effects being manifested. All the rest were cases of grave disease; and so far as the trial, which was interrupted about the fortieth day, extended, the patients grew worse, or received no benefit. A case is reported on the page before me of a soldier affected with acute inflammation in the chest, who took successively aconite, bryonia, nux vomica, and pulsatilla, [all popular homeopathic remedies, then and today] and after thirty-eight days of treatment remained without any important change in his disease.

The Homoeopathic physician who treated these patients was M. de Horatiis, who had the previous year been announcing his wonderful cures. And M. Esquirol asserted to the Academy of Medicine in 1835, that this M. de Horatiis, who is one of the prominent personages in the Examiner's Manifesto published in 1840, had subsequently renounced Homoeopathy. I may remark, by the way, that this same periodical, which is so very easy in explaining away the results of these trials, makes a mistake of only six years or a little more as to the time when this trial at Naples was instituted.

M. Andral, the "eminent and very enlightened allopathist" [orthodox physician] of the "Homoeopathic Examiner," made the following statement in March, 1835, to the Academy of Medicine:

> "I have submitted this doctrine to experiment; I can reckon at this time from one hundred and thirty to one hundred and forty cases, recorded with perfect fairness, in a great hospital, under the eye of numerous witnesses; to avoid every objection I obtained my remedies of M. Guibourt, who keeps a Homoeopathic pharmacy, and whose strict exactness is well known; the regimen has been scrupulously observed, and I obtained from the sisters attached to the hospital a special regimen, such as Hahnemann orders. I was told, however, some months since, that I had not been faithful to all the rules of the doctrine. I therefore took the trouble to begin again; I have studied the practice of the Parisian Homoeopathists, as I had studied their books, and I became convinced that they treated their patients as I had treated mine, and I affirm that I have been as rigorously exact in the treatment as any other person."
>
> And Andral expressly asserts the entire nullity of the influence of all the Homoeopathic remedies tried by him in modifying, so far as he could observe, the progress or termination of the diseases. It deserves notice that he experimented with the most lauded substances—cinchona, aconite, mercury, bryonia, belladonna. Aconite, for instance, he says he administered in more than forty cases of that collection of feverish symptoms in which it exerts so much power—according to Hahnemann—and in not one of them did it have the slightest influence, the pulse and heat remaining as before.

One could certainly expect that after such comprehensive and authoritative testing had resulted in total failure, homeopathy would immediately have vanished from the further consideration of the profession and the public. But to quote from Dr. Holmes (above): "to suppose that any trial can absolutely silence people, would be to forget the whole experience of the past." Homeopathy is still with us, and no doubt will survive any contrary evidence, simply because there is a huge commercial aspect to its continued existence, along with wide ignorance of how to judge these matters.

As powerful, comprehensive, and evidential as the BBC *Horizon* program was—and we're very happy that a major network has actually extended itself to do the testing procedure—history tells us that the homeopathic community, those with heavy financial and philosophical interests in supporting this quackery, will rally, regroup, and begin obfuscating wildly to neutralize this damning research. They certainly cannot deny those behind it: top-notch medical, biophysical, and biochemical authorities, using the very best experimental standards, and adopting a firm statistical conclusion. But they will squirm and mumble, wriggle and grumble, complaining that it just had to be something wrong with the experimental procedure, not the theory itself.

Here are a few samples of the more than 900 questions that arrived for me following the *Horizon* broadcast last Tuesday:

> If some labs are creating positive results for homoeopathy and it is shown that water does not have memory should we be worried that many labs are not rigorous enough to be looking after our health?
>
> Do you think that the results have anything to do with determinism? In a similar way to the Schrödinger's Cat scenario?
>
> Is there something wrong with science if it has to always prove how things work, not just that they work (repeatable observation of a phenomenon)?
>
> Do you personally believe in any pseudosciences?
>
> If homeopathy works, surely drinking one glass of water will cure me of everything, after all, it will have been in contact with most substances at some point. What do you think?

No comments on the above. . . .

Concerning the excellent handling of the homeopathic claims by *Horizon*, I repeat another caveat of Dr. Holmes: "realize the entire futility of attempting to silence this asserted science by the flattest and most peremptory results of experiment. Were all the hospital physicians of Europe and America to devote themselves, for the requisite period, to this sole pursuit, and were their results to be unanimous as to the total worthlessness of the whole system in practice, this slippery delusion would slide through their fingers without the slightest discomposure."

The transcript of the *Horizon* program can be seen at: http://www.bbc.co.uk/science/horizon/2002/homeopathytrans.shtml.

I urge you to read the entire Holmes account, available at www.quackwatch.org/01QuackeryRelatedTopics/holmes.html and get the whole matter in perspective. If you've not yet been exposed to pseudoscience at its very weirdest, you're in for a shock. This, friends, is what homeopathy is all about. . . .

## NOTE

**Originally published in *Skeptic* 10, no. 1 (2003): 6–8.**

# 13

## THE CASE FOR AND AGAINST ASTROLOGY

### *Geoffrey Dean*

> If I doubt astrology to a believer, I am looked at with a shocked and bewildered
> stare, as if I were attacking apple pie and motherhood.
>
> Anthony Standen, *Forget Your Sun Sign*, 1977

Astrology has been a field made quarrelsome by a shortage of facts. Could astrology be true? Could the stars really correlate with human affairs? Such questions have been furiously debated without resolution for more than two thousand years. Astrology has been the world's longest shouting match.

Not any more. Advances in related areas (astronomy, psychology, statistics, research design) and a decisive technology (personal computers) have since the 1970s put astrology under the scientific microscope like never before. A birth chart that could take hours or days to calculate by hand now takes less than a second. The result has been a minor explosion in empirical studies.

Before 1950 few empirical studies of astrology existed. Today there are hundreds, but many are not easily retrieved and are therefore rarely considered. The growth of research articles 1950–2004 is shown in Figure 1.

In Western languages the literature of astrology totals about one thousand shelf-feet of books and periodicals. On the Web, entering *astrology* into Google will return fifty million pages, or over a million for *astrology* just in the URL. Web-based book finders typically return over a thousand new and used astrology titles in English and in stock. But low standards prevail everywhere. Except for scholarly works, astrology books are characterised by unsupported and often contradictory assertions that are hard to take seriously. Disagreement is the rule, including disagreement on how to resolve disagreement. To debunk astrology you need only compare astrology books with each other.

However, most questions about astrology can now be answered. Quarrelling is no longer the option it once was. In what follows I will ignore the usual tired arguments against astrology (sun signs do not agree with the constellations, free will makes astrology nonfalsifiable, there is no known way astrology could work) in favour of meta-analyses of relevant studies involving hundreds of astrologers and thousands of birth charts. I will also look at recent developments involving Gauquelin's planetary effects. But first some views about astrology (the real thing, not newspaper horoscopes) from astrologers and scientists.

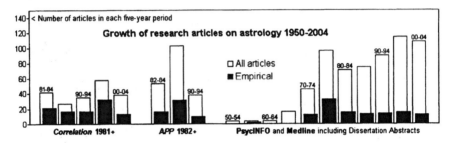

[**Figure 1.** Number of research articles on astrology published every five years in *Correlation*, the UK Astrological Association's journal of research into astrology, in the now defunct *Astro-Psychological Problems* published by Françoise Gauquelin, and in the journals abstracted by PsycINFO and Medline. The first two, founded in the early 1980s, are (or were) the leading research journals in English and can contain the better work but are not accessible via PsycINFO or Medline. Total articles are about 200, 200, and 600 respectively, of which empirical studies are about 100, 60, and 120. Non-empirical studies are usually a discussion or a historical essay. Empirical studies in other English and foreign astrological sources total about 250 but are often of poor quality.]

Astrologers leave you in no doubt that astrology works:

> Against its detractors, astrology will obtain an easy victory, a glorious triumph, by the force of its most powerful weapons—*facts* (Alfred John Pearce, for nearly fifty years the learned editor of Zadkiel's Almanac, in his first book *A Defence and Exposition of the Principles of Astrology*, 1863).

> Practical experiment will soon convince the most sceptical that the bodies of the solar system indicate, if they do not actually produce, changes in: 1. Our minds. 2. Our feelings and emotions. 3. Our physical bodies. 4. Our external affairs and relationships with the world at large (Charles Carter, leading British astrologer of his day, *The Principles of Astrology*, 1925).

> Official science will recognise that the ancients were not mistaken, and astrology, aided by new methods of investigation, will recover its ancient prestige (French astrologer Paul Choisnard, pioneer of empirical tests, *Les objections contre l'astrologie: Réponses aux critiques anciennes et modernes*, 1929).

> No one has ever been known to make a serious study of Astrology and then reject it (Nicholas de Vore, American astrologer and president of the now defunct Astrologic Research Society, *Encyclopedia of Astrology*, 1947).

From being an outcast from the fraternity of sciences, it seems destined to assume an almost central role in scientific thought. . . . its impact will be felt in the next twenty years (John Addey, leading British astrologer of his day whose opinion was based on his massive empirical research, *Astrology Reborn*, 1971).

Astrology throws light on every department of life; . . . From sex to career to character and future prospects—and more (American astrologer Sydney Omarr, whose astrology columns were then appearing in nearly three hundred newspapers, *Astrology's Revelations About You*, 1973).

There is no area of human existence to which astrology cannot be applied (Julia and Derek Parker, *The Compleat Astrologer*, 1975, which sold over a million copies in ten languages. The first is a former President of the Faculty of Astrological Studies, the UK's leading teaching and examining body).

Anyone who makes a serious and open-minded study of astrology becomes totally unable to scoff. Its truths are inarguable (Mary Coleman, Australian psychologist and astrologer, *Astro-Pick your Perfect Partner*, 1986).

[Astrology] despite the contemptuous guffaws of scientific orthodoxy, still continues to enthral the minds of some of our finest contemporary thinkers (Charles and Suzi Harvey, former President of the UK Astrological Association and former Editor of its journal, *Principles of Astrology*, 1999).

Astrology's symbols are the soul's language of life. They reveal not only the mysteries of the universe but also the mysteries of each of our lives (Gina Lake, American counselling psychologist and astrologer, *Symbols of the Soul: Discovering your Karma through Astrology*, 2000).

[Astrology] promises to contribute to the emergence of a new, genuinely integral world view, one that . . . can reunite the human and the cosmic, and restore transcendent meaning to both (Professor Richard Tarnas, American philosopher and astrologer, *Cosmos and Psyche*, 2006).

In short, astrology is all-revealing, factual, inarguably true, applicable to everything including past lives, enthralling to thinkers, soon to dominate scientific thought, the key to a new world view, and more. Just study it seriously and you will be convinced it works. Or so astrologers lead us to believe. Now a few words from scientists who have studied it seriously:

I myself . . . have for fourteen years held my archives open for astrological evidence, and have collected many exhibits of what was offered as evidence by supposed experts, . . . [but all were] the result either of a forced application of the rules to human careers already known, or of a careful culling of hits from preponderating numbers of misses. I do not think that any psychical researcher in forty-eight years has given attention to the claims of astrology and has not definitely cast the pretended science on the dust heap (Walter Franklin Price, presidential address to the Society for Psychical Research, 1930).

The ancients were evidently unaware that [astrological judgements] were the result of reasoning by analogy, which so often proves a treacherous foundation. That is why the whole superstructure of astrology is so utterly worthless and fallacious (August Thomen, *Doctors Don't Believe It*, 1938, a survey of medical superstitions hailed as the first critical look in three centuries).

The casting of horoscopes provides a living to thousands of individuals and provides dreams to an infinitely larger number of consumers. . . . [But] since the most painstaking studies have shown the inanity of horoscopes, there should be a strong rising up against this exploitation of public credulity (Michel Gauquelin, after analysing the birth charts of 16,000 famous people, *Dreams and Illusions of Astrology*, 1969).

The picture emerging suggests that astrology works, but seldom in the way or to the extent that it is said to work (Geoffrey Dean and Arthur Mather, *Recent Advances in Natal Astrology*, 1977, a critical review by fifty astrologers and scientists of over 1000 astrology books, 400 astrology journal articles, and 300 relevant scientific works that took seven man-years to prepare).

We are convinced however that astrology does not work. Astrology cannot be used to predict events of any kind, nor is astrology able to provide any useful information regarding personality, occupation, health, or any other human attribute (Roger Culver and Philip Ianna, *The Gemini Syndrome*, 1979, a review by two American astronomers of evidence collected over many years).

Astrology is largely (but not entirely) superstition. However, because of the important areas which remain to be investigated [matching tests, Mars effect], this conclusion may need future qualification. We should not be dogmatic (Hans Eysenck and David Nias, *Astrology: Science or Superstition?*, 1982, a review by two British psychologists of the latest scientific research).

Science isn't interested in astrology any more . . . no observable effect, no need to investigate any 'causes' (Daniel Fischer, editor-in-chief of *Astronomische Nachrichten*, in response to the question "Are there scientists trying to prove that astrology is true?" sent to the *Guardian* newspaper, London, 1994).

The single fact that astrologers contradict each other at about every point, and the firm convictions of their own correctness supported by their experience, must call up doubts about the reliability of [their] methods. . . . Not a single classical astrological element is shown to be able to resist statistical research (Ronnie Martens and Tim Trachet, *Making Sense of Astrology*, 1998, a critical review by two Belgian skeptics of astrological claims).

Evidently astrology works if studied by astrologers but not if studied by scientists. Hence the shouting match. But how is such disagreement possible? How can presumably rational people look at presumably the same evidence and reach such opposite conclusions? To find out we need to look at what is meant by "astrology" and "astrology works."

To different people astrology can mean quite different things. In Western countries there are four broad levels of meaning and interest as shown below with the rough percentage of the population involved at each level. The exact percentage varies with country but the difference between levels remains much the same. For comparison roughly 0.05 percent of the population are dentists.

**Level of interest**
1. Superficial—reads sun signs, seeks entertainment, 50%
2. Some knowledge—has own birth chart, seeks self-examination, 2%
3. Deep involvement—calculates charts, seeks meaning to life, 0.02%
4. Scientific—performs controlled tests, seeks answers, 0.00002%

On going through the levels there is a huge falling off in numbers and a dramatic change in what astrology means. At the first level are the readers of sun sign columns. They see astrology as entertainment. At the second level are those who have their birth chart (horoscope) calculated and read. They see astrology as an intriguing way of exploring themselves. At the third level are those who read birth charts to find meaning in their lives. They see astrology as a form of religion unconnected with the entertainment of sun sign columns. At the fourth level are those who test astrology scientifically. They see astrology as a popular belief worthy of study regardless of whether the belief is actually true. Overall there is a crucial change in focus from entertainment to religion and research. Clearly there is more to astrology than being true or false. So what is meant by "astrology works"? Can there be a simple answer? To explore this point we first need to put astrology to the test.

## Putting Astrology to the Test

Astrologers claim they can tell your character, abilities, health, love life, events, destiny, and more, just from your birth chart. It seems amazing that a handful of planets could show all this. A simple event (planet $X$ opposes planet $Y$) can co-occur with a complex event (birth of person $Z$), but it cannot prescribe its content because you cannot get more information out than was put in. Astrologers have the perfect answer—just try it! Put astrology to the test, they say, and you will be convinced it works. What could be more reasonable and more disarming of criticism? There are now hundreds of tests. What have they found?

In the 1970s it was found that extraversion scores tended to vary in accordance with sun sign claims, where odd-numbered signs (Aries onwards) are said to be extraverted and even-numbered signs (Taurus onwards) are said to be introverted. The effect replicated and was hailed by astrologers as scientific proof of their claims. But the effect size was tiny. The effect also disappeared when people unfamiliar with sun signs were tested, see Figure 2 overleaf. So it had a simple explanation—self-attribution due to prior knowledge of astrology. Ask Sagittarians (said to be sociable and outgoing) questions related to extraversion, such as whether they like going to parties, and astrology might tip their answer in favour of *yes* rather than *no*. The effect is an artifact. It looks like astrology but has a non-astrological explanation. Nevertheless the power of astrology to shift people's self-image, however slightly, deserves recognition.

Artifacts have always raged out of control in astrology. Some even became famous in their day as the best claimed evidence for astrology, for example John Nelson's correlation between planets and radio quality, Frank Brown's lunar effects on oysters, Carl Jung's astrological experiment with married couples, Vernon Clark's matching tests, and John Addey's harmonics. Such artifacts took decades to be uncovered. (See later for artifacts in Gauquelin's Mars effect.)

Take sets of birth charts jumbled up with descriptions of their owners. Can astrologers match birth charts to owners? In astrology books they do it all the time with unfailing success. But in the 54 studies to date, which involved over 740 astrologers and 1400 birth charts, the average success rate was no better than chance, see Figure 3 left. For these astrologers, many of them among the world's best, astrology performed no better than tossing a coin. In a further 18 studies involving over 650 clients and 2100 readings, clients were unable to pick

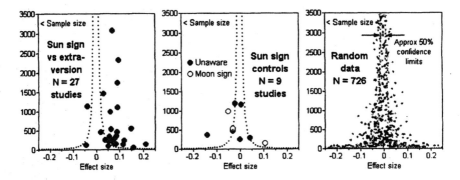

[**Figure 2**. Effect size as a correlation or similar measure for sun sign vs extraversion test scores (usually by EPI or EPQ). Dotted lines indicate the 50% confidence limits for effect size $r = 0.00$, approximated by $\pm 0.6745/\sqrt{n}$ where $\pm 0.6745$ is half the area under a normal curve and $n$ is the sample size. *Left:* The 27 studies to date show a clear positive deviation from $r = 0.00$. Meta-analysis by the bare bones procedure of Hunter & Schmidt (*Methods of Meta-Analysis: Correcting Error and Bias in Research Findings*, Sage 1990) gives a mean $r$ weighted by sample size of 0.070, *sd* 0.042, *p* 0.10. *Centre:* No deviation is observed when subjects are unaware of their sun sign or when their moon sign is tested (few people know their moon sign). Mean $r$ −0.020, *sd* 0.044, *p* 0.60. *Right:* How confidence limits work. If the true correlation $r$ is zero, as it will be for random data, then on average 50% of the observed values will lie within the 50% confidence limits.]

their own reading out of several, at least not when cues such as sun sign names or descriptions were absent, see Figure 3 centre.

To critics, astrology's failure to deliver is unremarkable because its alleged efficacy is explained by the same perceptual and cognitive biases that underlie proven invalid approaches such as phrenology and bloodletting. In the 1970s it was usual to explain client satisfaction as a Barnum effect, the reading of specifics into generalities such as "you have problems with money", where sense appears to come from the reading but in reality comes from our ability to make sense out of vagueness. Today more than thirty other "hidden persuaders" are known including cognitive dissonance (seeing what you believe), nonfalsifiability (nothing can count against your idea), the placebo effect (it does you good if you think it does), illusory correlation (finding meaning where none exists), selective memory (remembering only the hits), and the Dr Fox effect (blinding you with jargon as in this sentence). Each hidden persuader creates the illusion that astrology works, all are used routinely in consulting rooms, all lead to client satisfaction—and none require that astrology be true. But if clients are going to be satisfied, astrologers can hardly fail to believe in astrology. In this way a vicious circle of reinforcement is established whereby astrologers and clients become more and more persuaded that astrology works.

Astrologers predictably reject such views. They say test results are negative because the tests were too difficult or were made by people hostile to astrology (no matter that many were made by astrologers) or were not representative of what happens in their consulting rooms. They say you cannot test astrology by science, which if true would seem to deny their appeals to experience. Or they

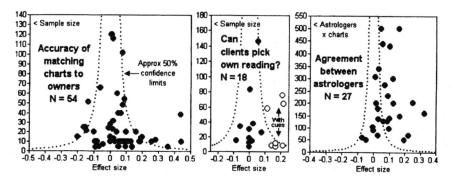

[**Figure 3**. Testing astrologers for their accuracy and agreement. *Left:* Effect size *r* for astrologers matching birth charts to personality, occupation, case histories, or their own questionnaires. The plot shows all known studies including unpublished ones and poor quality ones. Mean *r* 0.035, *sd* 0.117, *p* 0.77, equivalent to 51.75% hits when 50% is expected by chance. Observed effect sizes showed no relation to confidence or use of intuition. If astrology delivered as claimed in astrology books, *r* would be close to 1.00. Minimum *r* for tests generally accepted as being useful is around 0.4. Thus *r* for IQ vs achievement is around 0.6. *Centre:* When cues such as sun sign names or descriptions are absent from astrology readings, clients are unable to pick their own when given several. Mean *r* 0.020, *sd* 0.040, *p* 0.61, *N* 11. When cues are present the effect sizes show a positive deviation. Mean *r* 0.201, *sd* 0.125, *p* 0.11, *N* 7. *Right:* The agreement between astrologers shows a positive deviation. Mean *r* 0.098, *sd* 0.065, *p* 0.13. But none of the observed agreements come anywhere near the agreement of 0.8 generally recognised as being necessary for tests applied to individuals (as astrology is).]

see the bad news as proof of astrology's subtlety, so it is right even when it is wrong. The following test avoids such objections.

How well do astrologers agree on what a given birth chart indicates? To date 28 studies have put this to the test using in total more than 550 astrologers and 750 birth charts. Typically each test looked at how well 5 to 30 astrologers agreed on what a given chart indicated about its owner. Their average agreement was dismal, better than tossing a coin but nowhere near the minimum agreement acceptable for tests applied to individuals, see Figure 3 right. Again, many of these astrologers were among the world's best. But if astrologers cannot usefully agree on what a birth chart indicates, how can they know that astrology works? Indeed, why should anyone bother with astrology in the first place? It is here that we need to explore what is meant by "astrology works."

Originally astrologers claimed factual links between the heavens and human affairs, and such views continued through the 1950s. Then the rising interest in the inner person, and the rising frequency of disconfirming empirical studies, led to a retreat from factual claims to ones involving meaning. To most astrologers "astrology works" now means "astrology is meaningful." And astrology does indeed excel at being meaningful because it involves seeing faces in clouds of planetary gods. Because the ancient Greek founders of astrology chose gods that mirrored human conditions, the faces we see are our own, for example Mars *warlike*, Jupiter *benevolent*, Saturn *wise*. Astrology then guarantees a match with this celestial identikit by being unrelentingly flexible. Thus hard aspects can be simultaneously bad because their obstacles lead to failure and good because their challenges lead to success. If an awkward indication cannot be overturned by another factor, standard practice allows it to be explained away as untypical, or as unfulfilled potential, or as repressed, or as an error in the birth time, or as an outcome of the practitioner's fallibility. So it is always possible to fit any birth chart to any person, making it a most efficient focus for therapy by conversation. Once the astrologer and client are talking, the birth chart can mostly be ignored except as a convenient means of changing the subject.

Clearly this kind of astrology does not need to be true, and attacking it would be like attacking Superman comics or a religious faith. However, to critics (and also ambivalent astrologers when it suits them) astrology does need to be true, so that "astrology works" means "it delivers results beyond those due to non-astrological factors." In short, astrologers now officially judge astrology on how *helpful* it is, whereas critics have always judged it on how *true* it is. It is from this confusion that most shouting matches arise. The case for and against astrology can now be summarized.

## *The Case For and Against Astrology*

The case *for* astrology is that it is among the most enduring of human beliefs, it connects us with the cosmos and the totality of things, it provides a basic means of describing ourselves, and there is a wide range of approaches from the trivial to the profound. In emotional terms astrology delivers. In practical terms a warm and sympathetic astrologer provides non-threatening therapy by conversation that in a dehumanised society can be hard to find. You get emotional comfort, spiritual support, and interesting ideas to stimulate self-examination. In short, astrology seems to have its priorities right. The same applies to religion and pop psychology in all their forms, even though they often disagree and cannot all be right. Nobody can deny astrology's historical importance. And nobody who derives benefit from astrology is going to believe evidence for its invalidity.

The case *against* astrology is that it is untrue. It has not contributed to human knowledge, it claims the prestige of science without the methods of science, it has failed hundreds of tests, it does not deliver benefits beyond those produced by non-astrological factors (hidden persuaders), and users do not usefully agree on basics such as which zodiac to use or even on what a given birth chart indicates. No hint of these problems will be found in astrology books. Astrology also lends itself to commercial abuse as in sun sign columns and phonelines, and to exploitation of the gullible, which is why some critics see astrologers as misguided or even fraudulent. In fact most astrologers are nice people who genuinely wish to help others.

But the argument that astrologers repeatedly use (experience proves their claims) is simply mistaken—the experience they see as its strength is actually its weakness because the experience is not assessed under controlled conditions. Astrologers show little awareness of the factors such as the absence of accurate feedback that prevent learning from experience, or of the numerous hidden persuaders that give the illusion of such learning when it is actually absent. Astrologers also show little interest in procedures that avoid the weaknesses of experience, and every interest in ignoring unwelcome evidence, both classic hallmarks of a pseudoscience. Together these attitudes have created a case against astrology that is longer and stronger than the case for. Nevertheless both cases will be incomplete without a look at Gauquelin's findings.

## *Gauquelin's Findings*

Michel Gauquelin (1928–1991) was the world's most formidable scientific researcher into astrology, and his studies rank among the best ever conducted. Much of his success was due to Françoise (1929–), his Swiss-born wife and co-worker until 1985. His 45 years of research resulted in a dozen popular books (many translated into several languages), 30 thick data books, and about 150 scientific articles. He used rigorous methods and large samples of hundreds or thousands of cases. Among astrologers his positive results achieved the status of holy writ, defying (they said) every scientific explanation, and revealing man's true relationship with the cosmos. In 1979 my colleague Arthur Mather, in a now-famous quote, said it all:

> Both those who are for and against astrology (in the broadest sense) as a serious field for study recognise the importance of Gauquelin's work. It is probably not putting it too strongly to say that everything hangs on it (*Zetetic Scholar*, 3/4).

Gauquelin's interest in astrology began early. At fifteen he was skipping classes to browse in astrology bookshops, reading more than a hundred books in this way, and was busy writing his own. His chart readings were so successful that his schoolmates called him Nostradamus. But were the claims of astrology true? He was unconvinced by the success of his chart readings and by the opinions of astrologers, so he started collecting and testing birth data, an activity that was to continue for the rest of his life. In 1946 he enrolled at the Sorbonne to study psychology and statistics, graduating three years later with skills well beyond those of any previous researcher into astrology.

Gauquelin began by testing traditional claims ranging from simple ones such as zodiac signs vs personality to more complex ones such as transits at death and planetary aspects between family members. His findings were uniformly negative. In 1955 he published his results, stressing that they represent

> a considerable inquiry in the testing of astrological rules with large and varied samples. It is necessary to stress that the results demolish astrology more than they might appear . . . because they attack not the claims of particular authors but the elementary basis of the doctrine itself (*L'Influence des Astres*, 62).

The situation changed when Gauquelin tested the non-traditional findings of French astrologer Leon Lasson. In 1938 Lasson had published *Astrologie Mondiale . . .Quinze ans de paix sur l'Europe*, which predicted 15 years of peace in Europe, clearly not a promising start. But in a later work Lasson looked at how planets were distributed in the sky at the birth of 807 eminent professionals and

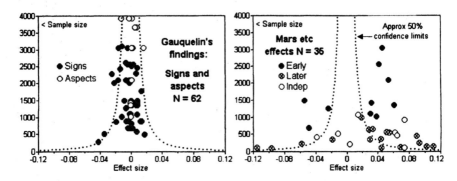

[**Figure 4**. Effect size vs sample size for the Gauquelin findings. *Left:* Gauquelin's 51 tests of signs (Sun, Moon, MC, and Ascending) and 11 tests of aspects vs the occupation or personality of eminent professionals as predicted by astrology. Mean $r$ 0.007, $sd$ 0.006, $p$ 0.29. Note that negative effect sizes (like Earthy farmers avoiding Water signs) cannot be counted as misses as in Figure.3, so what matters is the distribution of results with respect to the 50% confidence limits. Most results are within these limits because they are an average across relevant signs, which has reduced their scatter. *Right:* Effect sizes for planetary effects in Gauquelin's early studies of ten professional groups (often involving the same samples as on left), his 17 later studies, and eight independent studies including those by skeptics. The last two tend to involve smaller sample sizes reflecting the difficulty of finding new data. Mean $r$ 0.044, $sd$ 0.019, $p$ 0.02, equivalent to 52.2% hits when 50% is expected by chance. This time most of the results are beyond the 50% confidence limits, indicating that they are unlikely to have arisen by chance. Not shown are effect sizes for parent-children effects, typically about 0.02 (as phi) for $n$ = 16,000.]

claimed the results were significant. Gauquelin found artifacts that explained many of the results, but to his great surprise he found that the rest replicated and could not be explained by artifacts even though they were only partially related to traditional astrological claims. For example the favoured positions (nothing to do with signs or aspects) were traditionally weak instead of strong.

In due course Gauquelin found two things that pleased astrologers, upset critics, and puzzled everyone else. (1) Professional people such as scientists tended to be born with a surplus or deficit of certain planets in the areas just past rise or culmination (these were the traditionally weak positions), but only if the people were eminent and born naturally. (2) Ordinary people with such features tended to pass them on to their children. Both tendencies were too weak to be of practical use and often required thousands of births for their reliable detection. Figure 4 illustrates the marked difference in effect size between Gauquelin's negative and positive results. The difference had no obvious explanation.

The positive effect was later called the Mars effect because Mars was the significant planet for sports champions, who were then the focus of attention. But depending on the occupation (there were nine others) it could have been called the Moon, Venus, Jupiter or Saturn effect. There was no effect for the Sun, Mercury, Uranus, Neptune, or Pluto, or for ordinary people such as those who visit astrologers. The challenge of these findings lay less in the effects and more in the associated puzzles.

Ironically the Gauquelin planetary effects are as puzzling for astrology as they are for science. For science the puzzles include: Why no link with physical variables such as distance, why no link with the sun, why is eminence important, why an effect only at birth, why contrary to all expectation is the effect larger for rounded birth times, and why does it disappear when the birth is induced or surgically assisted?

For astrology the puzzles include: Why only rising/culminating positions and not signs or aspects, why occupation and not character, and why only five planets? After all, astrologers do not claim that astrology fails to work for half the planets, for signs, for aspects, for character, or (on Gauquelin's figures) for the 99.994 percent of the population who are not eminent.

The above puzzles seem utterly baffling. But might they be due to artifacts, something other than astrology that mimics planetary effects?

## Three Types of Artifact

Here three types of potential artifact can be identified: (1) Artifacts of *natural cycles* such as astronomy (Mars spends more time conjunct Sun than opposite Sun) and demography (birth rate varies with time and place). (2) Artifacts of *procedure* such as selection bias (use only data that work) and improper statistics (shoot enough arrows and some are bound to hit). (3) Artifacts of *social behaviour* such as faking by parents (adjust birth data to match a desired astrology) and self-attribution by subjects (adjust behaviour to match astrology).

Social artifacts may seem unlikely today, but most of Gauquelin's cases were born during the nineteenth and early twentieth centuries in Western Europe. It was a time when births were reported verbally to the registry office by the parents, when rising and sometimes culminating times for the visible planets were available in some popular almanacs, and when early world views still survived including the then pop tradition in astrology long established by *Le Grand Calendrier et Compost des Bergers* (The Big Calendar and Compilation of Shepherds), Western Europe's most popular almanac. *Le Compost* was first published in France in 1491, and its pop astrology section was being printed largely unchanged more than three hundred years later with equally long-lasting translations in other countries. It gave the occupations associated with each visible planet, for example Mars made you "s'adonne à faire guerre ou un grand chemineur par terre" (addicted to war or going on foot), so Mars was the planet of choice for parents with traditions in military service or competitive sport, and held that a body was strongest "quelle est en l'ascendant ou au milieu du ciel" (when above the horizon or culminating in the middle of the sky). In other words what people were reading matched what Gauquelin was observing.

Over the years the above potential artifacts have received much attention, often leading to bitter disputes in which critics tended to behave badly, but today the outcomes are clear. Natural and procedural artifacts (which Gauquelin had always recognised) do not explain most planetary puzzles, and planetary effects persist when the artifacts are controlled. So these artifacts remain artifacts, not an explanation of planetary puzzles. In contrast, social artifacts easily explain every puzzle including the apparent disappearance of planetary effects.

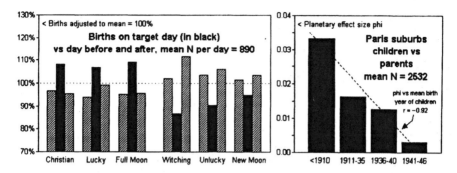

[**Figure 5.** Evidence for parental faking in the Gauquelin data. *Left:* If Gauquelin births occur at random then each distribution should be more or less uniform with no consistent trend. But births increase on days that European folklore says are desirable and decrease on days said to be undesirable, indicating that about 1 in 25 parents were faking the data they reported to registry offices. Furthermore, planetary effect sizes on desirable days (where faking is at a maximum) average twice those on undesirable days (where faking is at a minimum). Christian = 15 major fixed feast days such as Christmas and Epiphany. Lucky = 1st, 3rd, 7th of each month. Unlucky = 9th, 13th, last of each month. Witching = 10 witches' sabbats such as *Walpurgisnacht* 30 April. *Right:* Planetary links between parents and children tend to disappear as births become more recent and opportunities for parental faking decrease. Gauquelin was unaware of parental faking and suggested that the disappearance was due to recent births being increasingly subject to medical intervention, which in his view upset the natural planetary links, see text. But a decrease in faking is more plausible than all hospitals and all midwives intervening in all births. Also, the rebirth of serious astrology in the 1930s meant a gradual end to the popular ideas of *Le Compost* and its match with the Gauquelin findings.]

For example, Figure 5 illustrates the existence of parental faking in the Gauquelin data, and how it slowly disappeared as the verbal reporting of births was replaced by documents issued by the doctor or hospital. A direct connection between planets and parental faking is revealed by the cluster analysis shown in Figure 6. The planetary effect sizes in Figure 4 are roughly half those due to sun sign self-attribution in Figure 2, so a social basis is not implausible.

Indeed, social artifacts make some puzzles almost ridiculously easy to explain. Thus the planetary effect is limited to planets that could be seen in the sky or read in almanacs. It involved occupation and not character because that was the belief in those days. There is no effect for signs or aspects because in almanacs they were linked with the seasons or with weather. There is no link with physical variables such as distance or gravity because they were not part of popular belief. There is no effect for the sun because its position was relevant only to the seasons and to seasonal work on the farm. There is an effect only at birth because that was the popular belief. Occupation effects are strongest where family traditions are strongest, where the match between planets and occupation is closest, and where there is most need to be suitably destined, as in eminent families. Hence the importance of eminence.

But why is the planetary effect for precise birth times half that for birth times rounded to the hour? This is like saying the more we tune our radio the

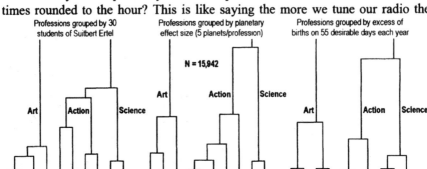

[**Figure 6.** Cluster analysis of Gauquelin's ten professional groups totalling 15,942 births. In 1990 Suitbert Ertel (in Tomassen et al., *Geo-Cosmic Relations*) had thirty students sort Gauquelin's ten professions into alike groups, for example painters and musicians seem more alike than painters and military men. Did their verdict match the planets' verdict? *Left:* the students identified three broad clusters that I have labelled Art, Action, and Science. *Centre:* the clusters based on planetary effect sizes are almost identical. Ertel concluded that planetary effects are not haphazard but conform to human views. *Right:* my cluster analysis of the excess births on 55 desirable days each year similar to those in Figure 5. The clusters are again almost identical, indicating that planetary effects are not haphazard but conform to parental faking. That is, their faking shows clustering in keeping with the planet of the associated births. For undesirable days the results are similar. Of course even random data will give clusters, and for effect sizes as tiny as these there will be much statistical uncertainty, so what matters is the number and content of broad clusters, not the order of components within them.]

worse the reception. It is not what we expect. No astrologer, no critic, not even Gauquelin would have predicted such a result. But faked birth times do not need to be precise. What matters is the planet's general position in the sky, not its exact position, so the precision to which clocks would normally be read is not needed. Rounded birth times are good enough, and as a bonus they do not raise suspicions at the registry office like an overly precise time might. In other words faking increases rounding and also increases planetary effects.

But why fake? If we really believe that certain times are auspicious, we can hardly believe that faking will change anything. On the other hand, if we do not believe, why bother? But look at it this way. Even if we saw nothing wrong with an inauspicious time, other people (and the child) might disagree, which could have dire consequences. So we fake, not because we necessarily believe but because others believe. Similarly, if we can fake an auspicious birth date, or a planetary indication of greatness in a chosen occupation, it could have useful consequences. Being suitably destined in the eyes of the child and others has advantages. The same motivation exists today when hotels omit 13 from floor and room numbers lest their occupancy be affected, and when psychologists control for the expectations of experimenters.

One final puzzle. Gauquelin had found that fathers and mothers contributed roughly equal amounts of planetary effect to their children, which finding he called his Rosetta Stone because it suggested a link with genetics. But he did not make same-sex and opposite-sex comparisons, and although he was aware of superstitious beliefs such as those favoring even hours over odd hours (which is actually detectable in his data with an effect size of about 0.02), he felt they were unlikely to simulate planetary effects. Blur, yes, simulate, no.

However, genetics predicts equal contributions regardless of sex whereas social artifacts predict unequal contributions. So we have the classic situation of two rival hypotheses. What is the evidence? Regardless of planet, the contribution by same-sex parents is roughly twice that of opposite-sex parents. In other words sons are more like their fathers and daughters are more like their mothers, confirming what Lady Catherine de Bourgh says in Jane Austen's *Pride and Prejudice* (1813), "daughters are never of much consequence to a father." Of the two rival hypotheses, the winner is social artifacts.

In summary, Gauquelin had found a real effect but contrary to what everyone thought there was no conflict with science and no need for disbelief. Data selection and fraud (a popular skeptical ploy) can be rejected because Gauquelin could hardly be selective or fraudulent about social artifacts he was unaware of. Nevertheless social artifacts are still artifacts, not an explanation of planetary effects. They are an explanation only if planetary effects disappear when the artifacts are controlled, as when births are reported by doctors instead of by parents, and the child is ignorant of its birth planets. As yet nobody knows for sure if they do or not, even though it could be argued that Gauquelin's failure to find planetary effects in births after 1950 has already put this point to the test.

## ACKNOWLEDGMENTS

I thank Ivan Kelly, Arthur Mather, and Rudolf Smit for helpful comments and for thirty years of unflagging critical assistance in investigating astrological claims.

## FURTHER READING

Progress during the 1990s was such that earlier reviews tend to be seriously incomplete. Expanded versions of the present article, references, and many other articles relevant to the scientific exploration of astrology can be found at www.astrology-and-science.com, which presents hard-to-find material in a very user-friendly way.

# 14

## LOOKING UP TO LOGIC

### *Bryan Farha*

Imagine you live in the small city of Salina, Kansas. As is typical of Kansas weather, it is a fairly windy morning. There is much debris blowing in your front yard, including leaves, branches, and twigs; the children's toys; and garbage resulting from the trash container having been blown over.

On the lawn of your residence, about twenty feet from your doorstep, you see what looks like a newspaper amongst all the debris. Because of the adverse conditions, it is somewhat difficult to *clearly* see the print on the paper. But you do not subscribe to the local newspaper. Upon further inspection, and to your absolute bewilderment, you seem to be able to make out the words *"London Times"* at the top. Also on page one, there appears to be a large photograph of the football team that won the Super Bowl the previous day. The images are a bit fuzzy, but it certainly seems as though the *London Times* is headlining the previous day's Super Bowl, which is not unusual for any paper. It seems rather bizarre, however, that an apparently current newspaper from another country would find its way to your lawn in Salina, Kansas. You do not, nor have you ever, subscribed to the *London Times*. Your curiosity and interest are escalating. So you go to your bedroom to put on a robe and slippers and then go outside in order to get a much closer look to determine if your eyes are deceiving you. But now, the "phantom" paper is gone. You check your neighbors' yards, but no paper is in sight.

How can this experience be explained? There are several possibilities. But the question must be addressed from two perspectives: (1) Was the perception, in fact, what the observer thought it was? (2) If the argument is plausible, under what conditions could it be accounted for?

Concerning the first perception, there are myriad factors to consider. Recalling the wind and resulting debris, was this actually a newspaper or could it have been wrapping paper, someone else's trash, or even reflected sunlight? If, in fact, it was a newspaper, can we *verify* that it was the *London Times*? That it was a photograph of the recent Super Bowl champions on the front page? What tangible evidence exists of your experience?

Concerning the second perception, how does a current issue of the *London Times* find its way to your residential lawn in Salina, Kansas? It is possible that the newspaper belonged to a neighbor and the wind blew it into your yard; this notion is easily supported or refuted with a small degree of legwork. But if this legwork does not yield a satisfactory explanation, we might then ask a series of other questions. If this, too, fails to provide adequate explanation, we then become faced with very tenuous possibilities. One such possibility is that a newspaper carrier from London, England, came to Salina, Kansas, by plane and delivered the paper to your house. Remote as this seems, it *is* a possibility. Not a perfectly logical or feasible answer, but the *possibility* is there.

It is at this point that drawing plausible conclusions based on *logic* becomes critical. Unfortunately, it is also at this point that much of the general public errs in drawing conclusions based on available evidence. A case in point: An object in the night sky is unidentified. This *does* make the object a UFO. But the term "UFO" only means that the object cannot be identified. If evidence is insufficient to ascertain its identity (or its reality), then the conclusions we can draw are very limited. Why, then, do so many jump to the conclusion that *if we can't identify the object, then it must be an alien spacecraft from another solar system or galaxy?* Understand, it might be an alien spacecraft, but before we can draw this conclusion we must have *substantial* evidence. Assuming that an unidentified object in the sky is an alien spacecraft is as tenuous and potentially erroneous as attributing the arrival of the English newspaper to Salina, Kansas, via a London carrier traveling by airplane. Yet in the newspaper example, we easily recognize the faulty thinking involved in making the assumption (hypostatic leap) of a Londoner making an unexplained home delivery to a nonsubscribing Kansas resident.

Why, then, does the UFO phenomenon seem to change the *thinking process* of so many? Is the UFO phenomenon, as well as other potentially anomalistic (paranormal) experiences, so intriguing that people allow it to alter their understanding of logical thinking? It is *possible* that alien spacecraft visit Earth, that abductions occur, that evangelists can reverse disease, that objects can move without apparent impetus, and that "ghosts" exist. But a word of caution: We cannot make positive conclusions about these phenomena without evidence, substantiation, and the use of logic. In other words, the *scientific method* must be employed as the basis for drawing conclusions regarding paranormal claims. Science is not a panacea for all explanation, but regarding paranormal claims it remains, by far, the best method. Let's not fall into the trap of abandoning science and logic because of curiosity and imagination. Rather, let's use curiosity and imagination as a springboard to the scientific method in order to draw accurate conclusions regarding mysteries of the universe.

NOTE

**Originally published in *Skeptical Inquirer* (January/February 1996): 55–56.**

# 15

## PRAYER AND HEALING:
## THE VERDICT IS IN AND THE RESULTS ARE NULL

### Michael Shermer

In a long-awaited comprehensive scientific study on the effects of intercessory prayer on the health and recovery of 1,802 patients undergoing coronary bypass surgery in six different hospitals, prayers offered by strangers had no effect. In fact, contrary to common belief, patients who knew they were being prayed for had a higher rate of post-operative complications such as abnormal heart rhythms, possibly the result of anxiety caused by learning that they were being prayed for and thus their condition was more serious than anticipated.

The study, which cost $2.4 million (most of which came from the John Templeton Foundation), was begun almost a decade ago and was directed by Harvard University Medical School cardiologist Dr. Herbert Benson and published in *The American Heart Journal*, was by far the most rigorous and comprehensive study on the effects of intercessory prayer on the health and recovery of patients ever conducted. In addition to the numerous methodological flaws in the previous research corrected for in the Benson study, Dr. Richard Sloan, a professor of behavioral medicine at Columbia and author of the forthcoming book, *Blind Faith: The Unholy Alliance of Religion and Medicine*, explained: "The problem with studying religion scientifically is that you do violence to the phenomenon by reducing it to basic elements that can be quantified, and that makes for bad science and bad religion."

The 1,802 patients were divided into three groups, two of which were prayed for by members of three congregations: St. Paul's Monastery in St. Paul, Minnesota; the Community of Teresian Carmelites in Worcester, Massachusetts; and Silent Unity, a Missouri prayer ministry near Kansas City. The prayers were allowed to pray in their own manner, but they were instructed to include the following phrase in their prayers: "for a successful surgery with a quick, healthy recovery and no complications." Prayers began the night before the surgery and continued daily for two weeks after. Half the prayer-recipient patients were told that they were being prayed for while the other half were told that they might or might not receive prayers. The researchers monitored the patients for 30 days after the operations.

Results showed no statistically significant differences between the prayed-for and non-prayed-for groups. Although the following findings were not statistically significant, 59% of patients who knew that they were being prayed for suffered complications, compared with 51% of those who were uncertain whether they were being prayed for or not; and 18% in the uninformed prayer group suffered major complications such as heart attack or stroke, compared with 13% in the group that received no prayers.

This study is particularly significant because Herbert Benson has long been sympathetic to the possibility that intercessory prayer can positively influence the health of patients. His team's rigorous methodologies overcame the numerous flaws that called into question previously published studies. The most commonly cited study in support of the connection between prayer and healing is: Randolph C. Byrd, "Positive Therapeutic Effects of Intercessory Prayer in a Coronary Care Unit Population," *Southern Medical Journal 81* (1998): 826–829. The two best studies on the methodological problems with prayer and healing include the following: Richard Sloan, E. Bagiella, and T. Powell. 1999. "Religion, Spirituality, and Medicine," *The Lancet*. Feb. 20, Vol. 353: 664–667; and, John T. Chibnall, Joseph M. Jeral, Michael Cerullo. 2001. "Experiments on Distant Intercessory Prayer." *Archives of Internal Medicine*, Nov. 26, Vol. 161: 2529–2536, www.archinternmed.com.

The most significant flaws in all such studies include the following:

*1. Fraud.* In 2001, the *Journal of Reproductive Medicine* published a study by three Columbia University researchers claiming that prayer for women undergoing in-vitro fertilization resulted in a pregnancy rate of 50%, double that of women who did not receive prayer. Media coverage was extensive. ABC News medical correspondent Dr. Timothy Johnson, for example, reported, "A new study on the power of prayer over pregnancy reports surprising results; but many physicians remain skeptical." One of those skeptics was a University of California Clinical Professor of Gynecology and Obstetrics named Bruce Flamm, who not only found numerous methodological errors in the experiment, but also discovered that one of the study's authors, Daniel Wirth (AKA "John Wayne Truelove"), is not an M.D., but an M.S. in parapsychology who has since been indicted on felony charges for mail fraud and theft, for which he pled guilty. The other two authors have refused comment, and after three years of inquires from Flamm the journal removed the study from its website and Columbia University launched an investigation.

*2. Lack of Controls.* Many of these studies failed to control for such intervening variables as age, sex, education, ethnicity, socioeconomic status, marital standing, degree of religiosity, and the fact that most religions have sanctions against such insalubrious behaviors as sexual promiscuity, alcohol and drug abuse, and smoking. When such variables are controlled for, the formerly significant results disappear. One study on recovery from hip surgery in elderly women failed to control for age; another study on church attendance and illness recovery did not consider that people in poorer health are less likely to attend church; a related study failed to control for levels of exercise.

*3. Outcome Differences.* In one of the most highly publicized studies of cardiac patients prayed for by born-again Christians, 29 outcome variables were measured but on only six did the prayed-for group show improvement. In related studies, different outcome measures were significant. To be meaningful, the same measures need to be significant across studies, because if enough outcomes are measured some will show significant correlations by chance.

*4. Selective Reporting.* In several studies on the relationship between religiosity and mortality (religious people allegedly live longer), a number of religious variables were used, but only those with significant correlations were reported. Meanwhile, other studies using the same religiosity variables found different correlations and, of course, only reported those. The rest were filed away in the drawer of non-significant findings. When all variables are factored in together, religiosity and mortality show no relationship.

*5. Operational Definitions.* When experimenting on the effects of prayer, what, precisely, is being studied? For example, what type of prayer is being employed? (Are Christian, Jewish, Muslim, Buddhist, Wiccan, and Shaman prayers equal?) Who or what is being prayed to? (Are God, Jesus, and a universal life force equivalent?) What is the length and frequency of the prayer? (Are two 10-minute prayers equal to one 20-minute prayer?) How many people are praying and does their status in the religion matter? (Is one priestly prayer identical to ten parishioner prayers?) Most prayer studies either lack such operational definitions, or there is no consistency across studies in such definitions.

*6. Theological Implications.* The ultimate fallacy of all such studies is theological. If God is omniscient and omnipotent, He should not need to be reminded or inveigled that someone needs healing. Scientific prayer makes God a celestial lab rat, leading to bad science and worse religion.

NOTE

**Originally published in *Skeptic* 12, no. 3 (2006).**

# 16

## BENNY HINN:
## HEALER OR HYPNOTIST?

### Joe Nickell

Benny Hinn tours the world with his "Miracle Crusade," drawing thousands to each service, with many hoping for a healing of body, mind, or spirit. A significant number seem rewarded and are brought onstage to pour out tearful testimonials. Then, seemingly by the Holy Spirit, they are knocked down at a mere touch or gesture from the charismatic evangelist. Although I had seen clips of Hinn's services on television, I decided to attend and witness his performance live when his crusade came to Buffalo, New York, last June 28–29. Donning a suitable garb and sporting a cane (left over from a 1997 accident in Spain), I limped into my seat at the HSBC Arena, downtown.

### Learning the Ropes

Benny Hinn was born in 1953, the son of an Armenian mother and Greek father. He grew up in Jaffa, Israel, "in a Greek Orthodox home" but was "taught by nuns at a Catholic school" (Hinn 1999, 8). Following the Six-Day War in 1967, he emigrated to Canada with his family. When he was nineteen he became a born-again Christian. Nearly two years later, in December 1973, he traveled by charter bus from Toronto to Pittsburgh to attend a "miracle service" by Pentecostal faith-healing evangelist Kathryn Kuhlman (1907–1976). At that service he had a profound religious experience, and that very night he was pulled from bed and "began to shake and vibrate all over" with the Holy Spirit (Hinn 1999, 8–14).

Before long Hinn began to conduct services sponsored by the Kathryn Kuhlman Foundation. Kuhlman died before Hinn could meet her personally but her influence on him was profound, as he acknowledged in a book, *Kathryn Kuhlman: Her Spiritual Legacy and Its Impact on My Life* (Hinn 1999). Eventually he began preaching elsewhere, including the Full Gospel Tabernacle in Orchard Park, New York (near Buffalo) and later at a church in Orlando,

Florida. By 1990 he was receiving national prominence from his book *Good Morning, Holy Spirit*, and in 1999 he moved his ministry headquarters to Dallas.

Lacking any biblical or other theological training, Hinn was soon criticized by other Christian ministries. One, Personal Freedom Outreach, labeled his teachings a "theological quagmire emanating from biblical misinterpretation and extra-biblical 'revelation knowledge.'" He admitted to *Christianity Today* magazine that he had erred theologically and vowed to make changes (Frame 1991), but he has continued to remain controversial. Nevertheless, according to a minister friend, "Outside of the Billy Graham crusade, he probably draws the largest crowd of any evangelist in America today" (Condren 2001).

Hinn's mentor, Kathryn Kuhlman, who performed in flowing white garments trimmed with gold (Spraggett 1971, 16), was apparently the inspiration for Hinn's trademark white suits and gold jewelry. From her he obviously learned the clever "shotgun" technique of faith-healing (also practiced by Pat Robertson and others). This involves announcing to an audience that certain healings are taking place, without specifying just who is being favored (Randi 1987, 228–229).

## Selection Process

In employing this technique, Hinn first sets the stage with mood music, leading the audience (as did Kuhlman) in a gentle rendering of

> He touched me, oh, He touched me,
> And, oh, the joy that filled my soul!
> Something happened and now I know
> He touched me, and made me whole. . . .

Spraggett (1971, 17) says that with Kuhlman, as it was sung over and over, it became "a chant, an incantation, hypnotic in its effect," and the same is true of Hinn's approach.

In time, the evangelist announces that miracles are taking place. At the service I attended, he declared that someone was being "healed of witchcraft"; others were having the "demon of suicide" driven out; still others were being cured of cancer. He named various diseases and conditions that were supposedly being alleviated and mentioned different areas of the anatomy—a back, a leg, etc.—that he claimed were being healed. He even stated that he need not name every disease or body part, that God's power was effecting a multitude of cures all over the arena.

Thus, instead of the afflicted being invited up *to be* healed (with no guarantee of success), the "shotgun" method encourages receptive, emotional individuals to believe they *are* healed. Only that self-selected group is invited to come forward and testify to their supposedly miraculous transformation. While I remained seated (seeing no investigative purpose to making a false testimonial), others are more tragically left behind. At one Hinn service a woman—hearing the evangelist's anonymously directed command to "stand up out of that wheelchair!"—struggled to do so for almost half an hour before finally sinking back, exhausted (Thomas 2001).

There is even a further step in the selection process: Of those who do make it down the aisles, only a very few will actually be invited on stage. They must first undergo what amounts to an audition for the privilege. Those who tell the most interesting stories and show the greatest enthusiasm are the ones likely to be chosen (Underdown 2001).

This selection process is—perhaps not surprisingly—virtually identical to that employed by professional stage hypnotists. According to Robert A. Baker, in his definitive book, *They Call It Hypnosis* (1990, 138–139):

> Stage hypnotists, like successful trial lawyers, have long known their most important task is to carefully pick their subjects-for the stage as for a jury-if they expect to win. Compliance is highly desirable, and to determine this ahead of time, the stage magician will usually give several test suggestions to those who volunteer to come up on the stage. Typically, he may ask the volunteers to clasp their hands together tightly and then suggest that the hands are stuck together so that they can't pull them apart. The stage hypnotist selects the candidates who go along with the suggestion and cannot get their hands apart until he tells them, "Now, it's okay to relax and separate them." If he has too many candidates from the first test, he may then give them a second test by suggesting they cannot open their mouths, move a limb, or open their eyes after closing them. Those volunteers who fail one or more of the tests are sent back to their seats, and those who pass all the tests are kept for the demonstration. Needless to say, not only are they compliant, cooperative, and suggestible, but most have already made up their minds in volunteering to help out and do exactly as they are told.

## *Role-playing*

Once on stage, one of Hinn's screeners announces each "healed" person in turn, giving a quick summary of the alleged miracle. At the service I attended, one woman put on a show of jumping up and down to demonstrate that she was free of pain following knee surgery three weeks before. Another was cured of "depression," caused by "the demon," said a screener, that resulted from "an abusive relationship with her husband." Still another (who admitted to being "an emotional person") said her sister-in-law sitting beside her had begun to "speak in tongues" and that she herself felt she was healed of various ailments, including high blood pressure and marital trouble. At her mention of her brother, Hinn brought him up and learned he had been healed of "sixteen demons" two years previously, and expected to be cured of diabetes; Hinn prayed for God to "set him free" of the disease. Another was supposedly cured of being "afraid of the Lord" (although he was carrying the bible of a friend who had died of AIDS), and one woman stated she believed she had just been cured of ovarian cancer.

In each instance-after the person has given a little performance (running about, offering a sobbing testimonial, etc.), and Hinn has responded with some mini-sermon, prayer, or other reaction-the next step in the role-playing is acted out. As one of his official catchers moves into place behind the person, Hinn gives a gesture, touch, or other signal. Most often, while squeezing the person's face between thumb and finger, he gives a little push, and down the compliant individual goes. Some slump; some stiffen and fall backward; a few reel. Once down, many lie as if entranced, while others writhe and seem almost possessed.

Along with speaking or praying in tongues (glossolalia) and other emotional expressions, this phenomenon of "going under the Power" is a characteristic of the modern charismatic movement (after the Greek *charisma*, "gift"). Also known as being "slain in the Spirit," it is often regarded skeptically even by other Christians who suspect-correctly-that the individuals involved are merely "predisposed to fall" (Benny Hinn: Pros & Cons 2002). That is, they merely engage in a form of role-playing that is prompted by their strong desire to receive divine power as well as by the influence of suggestion that they do so. Even the less emotionally suggestible people will be unwilling *not* to comply when those around them expect it.

In short, they behave just as if "hypnotized." Although popularly believed to involve a mystical "trance" state, hypnosis is in fact just compliant behavior in response to suggestions (Baker 1990, 286). One professional hypnotist said of Hinn's performance: "This is something we do every day and Mr. Hinn is a real professional" (Thomas 2001).

## Cures?

But what about the healings? Do faith-healers like Benny Hinn really help nudge God to work miracle cures? In fact, such claims are invariably based on negative evidence—"we don't know what caused the illness to abate, so it must have been supernatural"—and so represent the logical fallacy called "arguing from ignorance." In fact, as I explained to a reporter from *The Buffalo News* following a Benny Hinn service, people may feel they are healed due to several factors. In addition to the body's own natural healing mechanisms, there is the fact that some serious ailments, including certain types of cancer, are unpredictable and may undergo "spontaneous remission"—that is, may abate for a time or go away entirely. Other factors include even misdiagnosis (such as that of a supposedly "inoperable, malignant brain-stem tumor" that was actually due to a faulty CT scan [Randi 1987, 291–292]).

And then there are the powerful effects of suggestion. Not only psychosomatic illnesses (of which there is an impressive variety) but also those with distinct physical causes may respond to a greater or lesser degree to "mental medicine." Pain is especially responsive to suggestion. In the excitement of an evangelical revival, the reduction of pain due to the release of endorphins (pain-killing substances produced by the body) often causes people to believe and act as if they have been miraculously healed (Condren 2001; Nickell 1993; Nolen 1974).

Critical studies are illuminating. Dr. William A. Nolen, in his book *Healing: A Doctor in Search of a Miracle* (1974), followed up on several reported cases of healing from a Kathryn Kuhlman service but found no miracles—only remissions, psychosomatic diseases, and other explanations, including the power of suggestion.

More recently a study was conducted following a Benny Hinn crusade in Portland, Oregon, where seventy-six miracles were alleged. For an HBO television special, *A Question of Miracles* (Thomas 2001), Benny Hinn Ministries was asked to supply the names of as many of these as possible for investigation. After thirteen weeks, just five names were provided. Each case was followed for one year.

The first involved a grandmother who stated she had had "seven broken vertebras" but that the Lord had healed her at the evening service in Portland. In fact, x-rays afterward revealed otherwise, although the woman felt her pain had lessened.

The second case was that of a man who had suffered a logging accident ten years previously. He demonstrated improved mobility at the crusade, but his condition afterward deteriorated and "movement became so painful he could no longer dress himself." Yet he remained convinced he was healed and refused the medication and surgery his doctors insisted was necessary.

The next individual was a lady who, for fifty years, had only "thirty percent of her hearing" as claimed at the Portland crusade. However, her physician stated, "I do not think this was a miracle in any sense." He reported that the woman had had only a "very mild hearing loss" just two years before and that she had made "a normal recovery."

The fourth case was that of a girl who had not been "getting enough oxygen" but who claimed to have been healed at Hinn's service. In fact, since the crusade she "continued to suffer breathlessness," yet her mother was so convinced that a miracle had occurred that she did not continue to have her daughter seek medical care.

Finally, there was what the crusade billed as "a walking dead woman." She had had cancer throughout both lungs, but her doctors were now "overwhelmed" that she was "still alive and still breathing." Actually, her oncologist rejected all such claims, saying the woman had an "unpredictable form of cancer that was stable at the time of the crusade." Tragically, her condition subsequently deteriorated and she died just nine months afterward.

## *What Harm?*

As these cases demonstrate, there is a danger that people who believe themselves cured will forsake medical assistance that could bring them relief or even save their lives. Dr. Nolen (1974, 97–99) relates the tragic case of Mrs. Helen Sullivan who suffered from cancer that had spread to her vertebrae. Kathryn Kuhlman had her get out of her wheelchair, remove her back brace, and run across the stage repeatedly. The crowd applauded what they thought was a miracle, but the antics cost Mrs. Sullivan a collapsed vertebra. Four months after her "cure," she died.

Nolen (1974, 101) stated he did not think Miss Kuhlman a deliberate charlatan. She was, he said, ignorant of diseases and the effects of suggestion. But he suspected she had "trained herself to deny, emotionally and intellectually, anything that might threaten the validity of her ministry." The same may apply to Benny Hinn. One expert in mental states, Michael A. Persinger, a neuroscientist, suggests people like Hinn have fantasy-prone personalities (Thomas 2001). Indeed, the backgrounds of both Kuhlman and Hinn reveal many traits associated with fantasy-proneness, but it must be noted that being fantasy prone does not preclude also being deceptive and manipulative.

Hinn notes that only rarely does he lay hands on someone for healing, but he made an exception for one child whose case was being filmed for the HBO documentary. The boy was blind and dying from a brain tumor. "The Lord's going to touch you," Hinn promised. The child's parents believed and, although not wealthy, pledged $100 per month to the Benny Hinn Ministries. Subsequently, however, the child died.

Critics, like the Rev. Joseph C. Hough, President of New York's Union Theological Seminary, say of the desperately hopeful: "It breaks your heart to know that they are being deceived, because they genuinely are hoping and believing. And they'll leave there thinking that if they didn't get a miracle it's because they didn't believe." More pointedly, Rabbi Harold S. Kushner stated on *A Question of Miracles* (Thomas 2001):

> I hope there is a special place in Hell for people who try and enrich themselves on the suffering of others. To tantalize the blind, the lame, the dying, the afflicted, the terminally ill, to dangle hope before parents of a severely afflicted child, is an indescribably cruel thing to do, and to do it in the name of God, to do it in the name of religion, I think, is unforgivable.

Amen.

## ACKNOWLEDGMENTS

I am grateful to George E. Abaunza, Professor of Philosophy at Felician College in Lodi, N.J., for sending me a copy of the video *A Question of Miracles*. I also appreciate the input of Jim Underdown and other members of the Center for Inquiry—West's Independent Investigations Group who attended a Benny Hinn Miracle Crusade in Anaheim, California, August 17, 2001. Thanks are also due to Tim Binga for research assistance, Ranjit Sandhu for typing the manuscript, and Robert A. Baker for reading it.

## REFERENCES

**Originally published in *Skeptical Inquirer* 26, no. 3 (May/June 2002): 14–17.**

Baker, Robert A. 1990. *They Call It Hypnosis*. Buffalo, N.Y.: Prometheus Books.
Benny Hinn: Pros & cons. 2002. Internet posting:
    www.rapidnet.com/~jbeard/bdm/exposes/hinn/general.htm.
Condren, Dave. 2001. Evangelist Benny Hinn packs arena. The Buffalo News, June 29.
Frame, Randy. 1991. Best-selling author admits mistakes, vows changes. *Christianity Today*, October 28, 44–45.
Hinn, Benny. 1990. *Good Morning, Holy Spirit*. Nashville: Thomas Nelson.
———. 1999. *Kathryn Kuhlman: Her Spiritual Legacy and Its Impact on My Life*. Nashville: Thomas Nelson.
Nickell, Joe. 1993. *Looking for a Miracle: Weeping Icons, Relics, Stigmata, Visions & Healing Cures*. Amherst, N.Y.: Prometheus Books.
Nolen, William A. 1974. *Healing: A Doctor in Search of a Miracle*. New York: Random House.
Randi, James. 1987. *The Faith Healers*. Buffalo, N.Y.: Prometheus Books 228–229.
Spraggett, Allen. 1971. *Kathryn Kuhlman: The Woman Who Believed in Miracles*. New York: Signet.
Thomas, Antony. 2001. *A Question of Miracles*. HBO special, April 15.
Underdown, James. 2001. Personal communication, October 23.

# 17

## JOHN OF GOD: ABC'S *PRIMETIME LIVE* SPECIAL ABOUT A BRAZILIAN "HEALER" FALLS FAR SHORT OF INVESTIGATIVE JOURNALISM

### *James Randi*

On February 10th, 2005, ABC's *Primetime Live* aired an hour-long special on a man called "John of God" who works out of Brazil. This was a major show that could have been a useful, productive, and informative program, but failed to reach that standard.

The show dealt with the Brazilian "healer" João Teixeira, known popularly in his country as Joao de Deus, or "John of God." He holds forth at the "Casa de Dom Inacio," a healing center in Abadiania, Brazil, a tiny town about 70 miles southwest of the capital, Brasilia. Desperately ill people from all over the world flock to this place seeking cures. There is a group of thriving agencies organized to perpetuate the mythology surrounding the man, booking hotels, selling tours, trinkets, charms, and every sort of material that will satisfy these vulnerable people's hope for healing. The tours can run into the thousands of dollars, though tour operators are coy about actually naming prices. Their instructions to would-be visitors who apply for visas, is to not mention that they re going to Brazil for this purpose, because the government does not encourage such trips. The hotel stays start at two weeks and can run longer.

### *The Seven Stunts of John of God*

What brings these victims in from far and wide? There are seven stunts and other anomalies that grab public and media attention:

1. The forceps-in-the-nose trick.
2. The random cutting of the flesh.
3. The "scraping" of the eyeball.
4. The absence of immediate pain as a result of numbers 1, 2, and 3.
5. The subsequent absence of infection.
6. The "trances" John of God enters into in order to "contact spirits."
7. The subsequent recoveries reported by patients.

On January 25th, 2005, I was filmed by ABC television for this special, to offer my opinions and observations. They indicated that they wanted a skeptical point of view on these "miracles," and I had already seen a videotape prepared and distributed by the organization in Brazil that touted Teixeira's abilities. Little evidential information was in the video except that some of the common tricks and misleading claims made by the Casa de Dom Inacio organization were there to be clearly seen and revealed. ABC-TV also invited Dr. Mehmet Oz, a cardiac surgeon with the Columbia University Department of Surgery in New York, to be interviewed with me by their on-camera host, John Quiñones.

## The Wizard of Oz

I traveled to NYC, with my expenses (but no fee or other payment) covered by ABC-TV, and showed up at the recording studio. In the waiting area, I had the opportunity for a long conversation with Dr. Oz, and we came to general agreement on some major points—such as the matters of pain and infection (items 4 & 5 above)—that were sure to come up in the videotaping. However, I also discovered during that conversation that Dr. Oz takes a rather fanciful view of the real world, in spite of his very down-to-earth profession. He supports, and has written extensively on, such "complementary therapies" as hypnosis, "therapeutic touch," guided imagery, reflexology, aromatherapy, prayer, yoga, and "energy medicine," and he encourages their use 'in combination with the latest surgical techniques." Dr. Oz believes in using "alternative" techniques— which he refers to as "Global Medicine"—in an attempt to unblock and balance the chakras (energy channels), thereby boosting the body's natural healing capacities. His "complementary care" team at Columbia follows up with studies using Kirlian photography. All these are totally quack notions, and I began to suspect why he had been chosen by ABC-TV News as a participant.

For example, Dr. Oz said on the show: "Crawfish re-grow their nerves, right? Maybe there are things that we could harvest in our psyche that allow us to do it as well." The scientific approach on such a claim would be to *establish that the phenomenon itself exists in humans before you offer theories on how it works.* Oz also told viewers that the "visible surgery" performed by Teixeira "could be an old magician's trick, but it's a pretty powerful one from a physician's perspective." Oz learned about that trick from me during our pre-taping off-camera conversation; and what does he mean by "powerful"? Powerful enough to bring in the suckers to be "healed" by John of God? Dr. Oz also offered this comment on the forceps-in-the-nose stunt: "I'm wondering if touching the pituitary gland may influence all those chemicals that go between the body and brain."

Incredible! Dr. Oz is groping around to save this carny stunt and John of God by invoking such a remote, unlikely notion—that the forceps this charlatan pokes into the victim's nostril might touch the pituitary gland and thereby produce some actual effect. No, strike that. It's not "unlikely," it's *impossible.* The round-shaped pituitary gland would be accessible through the nasal passage only after an instrument went *about 5.5 inches into the nostril* (the forceps can only go in about four inches) and *through two thick layers of skull bone!* Again, Dr. Oz is postulating theories on a phenomenon that doesn't exist! There is zero evidence to show that John of God has ever accomplished anything but revulsion by sticking forceps up a victim's nose!

## The Taping Session at ABC News

The videotaping began. I sat before the ABC-TV cameras alongside Dr. Oz, prepared to enter into a dialogue with the doctor and respond to questions from the host John Quiñones. We recorded for about an hour, commenting on pertinent video material from Brazil that was shown to us on a studio monitor.

The result of this recording session, inserted into the broadcast, was hardly what I'd expected. I was introduced as "a debunker of the paranormal" with no other attribution. The opening phrase of the statement I made in my 19-second appearance wasn't even my own. The video recording was edited so that it appeared that these were my own spontaneous words: "There are no greater liars in the world than quacks—except for their patients." But *these were not my words.* I clearly said, preceding that quotation, "To quote Ben Franklin . . ." but it appears that ABC chose to put those words in my mouth to demonstrate what a cantankerous old curmudgeon I am, that I simply chose to rail against John of God as a quack. This seemed to indicate that I had nothing to contribute to the program except to rudely label John of God, and to scoff at him without adding anything to the discussion, while not providing any expert information. The rest of my statement was as follows:

Remember, these people have gone there for this kind of bizarre treatment. If they have to admit, "No. I'm not helped by this, I was swindled," they have to say, "I was pretty damn stupid to go in there and think that sticking something up my nose was going to cure my back!"

## *Why Patients Support the Scam*

I said this to illustrate the truth of Ben Franklin's observation that those who unwisely fall for such scams are often the fiercest defenders of their initial decision to become involved, and they choose to endorse the swindle even after it has failed them, rationalizing and even hyperbolizing to bolster their conviction. We must remember, however, that such people are desperate and thus vulnerable, both aspects that brought them there in the first place.

I had plenty more to say on the subject, *and I said it.* But ABC didn't use it. If they had, viewers would not only have been better-informed, but would have understood the true nature of the information they were being shown. Look, I'm very much aware of the fact that most of an interview can fail to be included in a final edited product, but much of what I provided for ABC-TV to use was pertinent data for achieving clarity on a controversial and critically important subject. Dr. Oz, who knows nothing about possible trickery, appeared in six lengthy inserts, offering what were, in my opinion, the appropriately woo-woo phrases that ABC-TV preferred on screen.

Note, too, that the John of God organization has set up a situation in which they simply cannot fail; if recovery is not experienced by their victims, it's not a failure of the magical forces, but the fault of the patient. They state that sometimes a person comes to them for a healing "too late," so it doesn't happen. If a patient doesn't have "the right attitude," or doesn't "keep the faith," the healing will fail. If the rules aren't followed no healing will occur. They say that one has to wait at least 40 days to see any healing—well after the victim has left Brazil—and sometimes up to two years have to pass before any effect will be seen. All this is a fail-safe scenario, one I've come upon many times in the faith-healing racket.

## *Forceps-up-the-Nose Stunt*

In a dramatic scene in the show, John of God inserts a forceps up a patient's nose. I explained on camera that this is an old carny effect that my friend Todd Robbins tells me dates back to the *jaduwallahs of* India and was adopted from their repertoire by an American performer named Melvin Burkhardt, first being done on this continent in 1926. It's now known as the "Blockhead Trick," and is usually done with a heavy 4 1/2" (thirty-penny) iron nail tapped up the nose and into the back of the throat, a clear, straight path that seems improbable. It is done today by more than 100 performers in carnivals and sideshows around the world, and John of God simply uses it to impress his victims, though he has a far easier time of it by using smooth nickel-plated (or stainless-steel) forceps.

I obtained from Todd Robbins a videotape of him doing the trick, and I look it to the ABC producer in New York, who chose not to use it even though it was *clearly* the trick used by John of God. ABC decided not to inform their audience that this impressive "miracle" is only a common carnival stunt, nor did they make clear that it is very simply explained and not in any way supernatural!

On-camera host John Quiñones told the audience that he had an inflamed rotator-cuff problem in his right shoulder and had submitted himself to treatment by John of God as a "test" of his powers. He was told by the Casa de Dom Inacio handlers to submit to the "invisible surgery," which consisted of merely meditating for two days and following a set of simple instructions— no sex, no pork, no alcohol, and no pepper— and then waiting *40 days* to see the results. Quiñones reported no change in his condition at all, but excused that failure by revealing that he'd not followed the instructions! Why was it that this professional investigative reporter, actively at work on a major media shoot looking into the claims of a controversial healer, chose not to follow the instructions he was given, thus providing a convenient excuse for the failure of the 'magic? And why, knowing that Quiñones had made his own ^TI test invalid by violating the rules, did the ABC editors and producers *still choose to include that event* in the program?

There are two kinds of "operations" performed by John of God, "visible" and "invisible." The "visible" ones are the forceps-up-the-nose, the reckless random slashing of the flesh, and the maneuver of the knife on-the-eyeball. The "invisible" ones consist of prayer, meditation, reading of holy scriptures, and sitting with eyes closed. Why was it that Quiñones was not "visibly" operated on? Could it be that the "healer" was smart enough not to give any actual physical distress to this representative of a powerful American media outlet that could give him—and *did* give him—priceless publicity and validation?

I've performed the famous 'psychic surgery" stunt many times, all over the world, notably on the *Tonight Show* with the late Johnny Carson many years ago. It consists of the performer apparently reaching into the body of the person on the table and extracting bloody lumps said to be tumors. That was clearly presented as a trick, and I explained on the *Tonight Show* that it was done by exactly the same means that the fakers in the Philippines are still using to cheat their victims. Imagine my surprise when Fred DeCordova, the *Tonight Show* producer, called to tell me that at the NBC-TV Las Angeles office alone, they'd received 102 phone calls following the broadcast, *every one of them asking how to contact the Filipino "psychic surgeon!"*!

Well. I believe that this ABC-TV program will—even more than my *Tonight Show* appearance— encourage the incautious public to book trips to Brazil to go under the butchery that John of God inflicts on his victims. The huge difference here is that while the Filipino "surgeons" seldom if ever actually break the skins of their customers, the Brazilian faker regularly does so, and that means not only financial loss, but injury and very possibly loss of life, as well. Is ABC willing to accept the grief and damage that those misinformed people will suffer?

## The Eyeball Scraping

John of God will seat a subject for his "visible surgery" stunt and apparently scrape the eyeball of the patient with the edge of a knife. I believe that this is a variation of the usual trick—illustrated on page 177 of my book *Flim-Flam!*—in which a knife-blade is inserted under the eyelid of a subject with little or no resulting discomfort. With the Brazilian faker, the "scraping" motion gives it a much more fearsome aspect, but for several good reasons I doubt that any contact takes place with the cornea.

The sclera (the white section of the eye) is relatively insensitive to touch. Try touching that area with a finger or any dean object, and you'll see this is true. The cornea, however, is very sensitive—among the most sensitive areas of the body. Mast persons (and I'm one of them) have a difficult time watching the eye being touched. We tend to empathize with the situation, and I'm sure that some readers are at this moment involuntarily squinting in distaste as they read these words. Few persons will resist looking away when John of God seems to scrape an eyeball, and I note that he was furtively watching the position of the camera as he performed this stunt, blocking the view with his body when a close-up was sought.

There is also the distinct possibility here that John of God introduces a temporary local anesthetic—benzocaine would work—onto the eye surface, which would allow contact with the cornea. Unless an anesthetic has been introduced, it is impossible for this man to be touching the cornea of a human eye as he appears to do without causing immediate involuntary flinching from the patient. The JREF will stake its million-dollar prize on that statement.

## Adrenaline Rush and Pain

An adrenaline rush is often experienced by people who are under unaccustomed stress or sudden shock. We're all familiar with accounts of soldiers in battle who are wounded in ways that would otherwise cause them great pain and bring an immediate reaction, but they remain unaware of the injury until the stressful conditions are relaxed. I've seen this happen when people onstage in front of a faith-healer's audience manage to do things that would have otherwise brought great discomfort and pain. In the case of John of God's sudden incisions, and considering the relatively insensitive areas he chooses to make these cuts, along with the fact that the victims are told to keep their eyes closed, I'm not surprised at the fact that—so far as the Casa de Dom Inacio people will permit us to see!—the victims show little or no reaction to the cutting procedure. But remember that we are only allowed to see the incision, not the possible subsequent reaction after the cameras are taken away.

## The "Trance State" and "Spirits"

Where was there *any evidence at all* in the ABC broadcast that a "trance state" of any sort was present with John of God? The "healer" claims that he's "taken over" by spirits of long-dead medical doctors. Host Quiñones expressed no doubts about this claim, which is so easily accepted and believed by many Brazilians because of their cultural traditions. No probing questions about it were asked, no proof was looked for—it was all allowed to slide by uncritically. Incredibly, John of God piously claims that when his body is taken over by any one of 40 different deceased doctors and/or King Solomon (?), he becomes unaware of what he does or says (surely the handiest "out" that could be imagined). And, when interviewer Quiñones bravely suggested that John might be getting rich at this practice, the ABC-TV cameras moved into an excruciating close-up of the beleaguered man with trembling lip and tear-filled eyes, asserting that he gave away all his money to the poor. As another reputable interviewer at ABC likes to say. "Give me a break!"

## The Absence of Infection

During my on-camera discussion with Dr. Oz, I was challenged to explain why the patients of John of God who undergo actual invasion of their bodies, suffer no infection—another "miracle" claimed by John of God. I turned to Dr. Oz and asked him if I was correct in stating that not all breaking of the skin—incisions, scrapes, punctures— resulted in invasion and proliferation by bacteria or viruses, and that it should not be assumed that an unsterilized instrument always brought on infection. Dr. Oz agreed. That comment and discussion was not used by the editors, even though one of the most powerful arguments used to support the supernatural nature of the operations in Brazil is this "non-septic" factor! But including this basic biological fact could have spoiled a perfectly good sensational story.

## An Appeal to Reason and Common Sense

Let's think about the whole picture here. In sending a film crew off to Brazil on this project, ABC took no prior advice from expert investigators, but actually seriously considered that these claims were possibly viable:

1. The body of a man in Brazil is periodically "occupied" by "beings of light"—spirits of an assortment of dead doctors—including King Solomon—who provide him with their accumulated medical expertise.

2. This man can glance at a person and correctly diagnose their disease(s).

3. He can heal a breast tumor by sticking a surgical clamp up the patient's nose, and a nervous condition by passing a knife over the patient's eyeball.

4. He goes into trances and has no memory of what he does when he performs his stunts.

Where, in the wide assortment of fantastic possibilities that ABC-TV might accept as suitable subjects to present seriously to the American public, do we place Santa Claus and flying pigs? If they had a report that a fat man in a red suit was manufacturing toys at the North Pole, would they dispatch a video crew to check it out? No, I think not, because making a small effort to contact experts would doom that project; but they eagerly rode off pursuing the four ridiculous premises listed above! And, I hasten to add—though they did not seek my advice on this matter—pigs cannot fly.

## How ABC Should Have Covered This Story

When you discover that the healings you're looking for *are not there,* and all you have are anecdotal accounts supplied by the organization you're investigating, the proper conclusion is that the claims are spurious—that you've been lied to, as the public you serve has been lied to. Is that too radical an approach? I think not.

I am not saying that John of God should not have been investigated and used as a subject for an ABC special, because this is a major matter of public curiosity, and it is news. This bizarre circus needs to be looked at and reduced to facts, with the fantasy aspects properly identified. What I am saying is that the project should have been done better so that lives might be saved and the public could be more properly informed. I dare to say that if ABC's John Stossel had been involved in the production of this program, the results would certainly have been far different. John would have researched the subject, might have called on me or another experienced observer for consultation *in advance of obtaining the raw material,* and would have used more reason and common sense.

Though I earnestly wish it could be different, based on what we *know* to be the hard facts, for those featured in the ABC special, David Ames will not recover from ALS (amyotrophic lateral sclerosis), Lisa Melman will most probably die of breast cancer because she's decided to forego legitimate surgical help, Matthew Ireland's brain tumor will still be there and will probably kill him, and João Teixeira will continue to flourish and be worshiped.

## What's the Harm?

I was once in Mexico City on the plaza outside the shrine of the Virgin of Guadalupe when a young peasant father crawled by me along the rough pavement with an obviously dead infant in his arms, swaddled in a tiny white serape. There were twin tracks of blood behind him from his bleeding knees. He was seeking a miracle. Through the adjacent barred window in the basilica I could hear the coin-sorting machines packaging the money that was pouring into the offering boxes inside. I turned away and wept.

In a St. Louis auditorium I stood in the lobby as paramedics treated a heavy elderly woman who lay in a fetal position on the carpet, white-faced and moaning in agony. Moments before she'd been seized in ecstasy in front of faith healer "Reverend" W. V. Grant, leaping up and down in an adrenaline rush that made her temporarily oblivious to the bone spurs on her arthritic spine that were cutting into her muscle tissues and bringing about internal bleeding. The attendants got her onto two stretchers and into an ambulance. I wept.

Outside an arena in Anaheim, California, my camera crew approached a tiny, thin, Asian boy with twisted legs on worn crutches to ask him if he'd been healed by Peter Popoff, the miracle-worker who, he'd told us two hours earlier, was "gonna ask Jesus to fix my legs." When he turned toward us, we saw his tear-streaked face and anguished eyes. The cameraman lowered his camera. "I can't do this," he said, and we both turned away and wept.

I have had my share of tears and sleepless nights, wondering what I might do to keep people from chasing this chimera. I had another chance in New York City with ABC-TV on January 25th, 2005, and I tried. I'd like to suggest to the government of Brazil that they shut down this charlatan and stop the victimizing of innocent- but-naive people from all over the world. The international community can only look in astonishment upon any nation that allows such flummery in the twenty-first century. I'm well aware that Brazil is far from being the only country plagued by such a burden; here in the U.S. we have Benny Hinn, W. V. Grant, Peter Popoff, and many others who perform the same level of fakery on our citizens. The U.K. is full of similar scams, and all over Europe we find the operators at work.

Fifteen weeks ago, I sent letters to David Westin, president of the ABC-TV News division, ABC News VPs Kerry Marash, Phyllis McGrady and Paul Slavin, and producer Shelley Ross. I told them the whole story as outlined here. I have never even received an acknowledgement of any of those letters. ABC-TV simply doesn't care—because they don't have to. . . .

NOTE

**Originally published in *Skeptic* 11, no. 4 (2005): 6–11.**

# 18

## BIGFOOT AT 50:
## EVALUATING A HALF-CENTURY
## OF BIGFOOT EVIDENCE

### Benjamin Radford

Though sightings of the North American Bigfoot date back to the 1830s (Bord 1982), interest in Bigfoot grew rapidly during the second half of the twentieth century. This was spurred on by many magazine articles of the time, most seminally a December 1959 *True* magazine article describing the discovery of large, mysterious footprints the year before in Bluff Creek, California.

A half century later, the question of Bigfoot's existence remains open. Bigfoot is still sought, the pursuit kept alive by a steady stream of sightings, occasional photos or footprint finds, and sporadic media coverage. But what evidence has been gathered over the course of fifty years? And what conclusions can we draw from that evidence?

Most Bigfoot investigators favor one theory of Bigfoot's origin or existence and stake their reputations on it, sniping at others who don't share their views. Many times, what one investigator sees as clear evidence of Bigfoot another will dismiss out of hand. In July 2000, curious tracks were found on the Lower Hoh Indian Reservation in Washington state. Bigfoot tracker Cliff Crook claimed that the footprints were "for sure a Bigfoot," though Jeffrey Meldrum, an associate professor of biological sciences at Idaho State University (and member of the Bigfoot Field Research Organization, BFRO) decided that there was not enough evidence to pursue the matter (Big Disagreement Afoot 2000). A set of tracks found in Oregon's Blue Mountains have also been the source of controversy within the community. Grover Krantz maintains that they constitute among the best evidence for Bigfoot, yet longtime researcher Rene Dahinden claimed that "any village idiot can see [they] are fake, one hundred percent fake" (Dennett 1994).

And while many Bigfoot researchers stand by the famous 16 mm Patterson film (showing a large manlike creature crossing a clearing) as genuine (including Dahinden, who shared the film's copyright), others including Crook join skeptics in calling it a hoax. In 1999, Crook found what he claims is evidence in the film of a bell-shaped fastener on the hip of the alleged Bigfoot, evidence that he suggests may be holding the ape costume in place (Dahinden claimed the object is matted feces) (Hubbell 1999).

Regardless of which theories researchers subscribe to, the question of Bigfoot's existence comes down to evidence- and there is plenty of it. Indeed, there are reams of documents about Bigfoot-filing cabinets overflowing with thousands of sighting reports, analyses, and theories. Photographs have been taken of everything from the alleged creature to odd tracks left in snow to twisted branches. Collections exist of dozens or hundreds of footprint casts from all over North America. There is indeed no shortage of evidence.

The important criterion, however, is not the *quantity* of the evidence, but the *quality* of it. Lots of poor quality evidence does not add up to strong evidence, just as many cups of weak coffee cannot be combined into a strong cup of coffee.

Bigfoot evidence can be broken down into four general types: eyewitness sightings, footprints, recordings, and somatic samples (hair, blood, etc.). Some researchers (notably Loren Coleman 1999) also place substantial emphasis on folklore and indigenous legends. The theories and controversies within each category are too complex and detailed to go into here. I present merely a brief overview and short discussion of each; anyone interested in the details is encouraged to look further.

## Eyewitness Accounts

Eyewitness accounts and anecdotes comprise the bulk of Bigfoot evidence. This sort of evidence is also the weakest. Lawyers, judges, and psychologists are well aware that eyewitness testimony is notoriously unreliable. As Ben Roesch, editor of *The Cryptozoological Review*, noted in an article in *Fortean Times*, "Cryptozoology is based largely on anecdotal evidence. . . . [W]hile physical phenomena can be tested and systematically evaluated by science, anecdotes cannot, as they are neither physical nor regulated in content or form. Because of this, anecdotes are not reproducible, and are thus untestable; since they cannot be tested, they are not falsifiable and are not part of the scientific process. . . . Also, reports usually take place in uncontrolled settings and are made by untrained, varied observers. People are generally poor eyewitnesses, and can mistake known animals for supposed cryptids [unknown animals] or poorly recall details of their sighting. . . . Simply put, eyewitness testimony is poor evidence" (Roesch 2001).

Bigfoot investigators acknowledge that lay eyewitnesses can be mistaken, but counter that expert testimony should be given much more weight. Consider Coleman's (1999) passage reflecting on expert eyewitness testimony: "[E]ven those scientists who have seen the creatures with their own eyes have been reluctant to come to terms with their observations in a scientific manner." As an example he gives the account of "mycologist Gary Samuels" and his brief sighting of a large primate in the forest of Guyana. The implication is that this exacting man of science accurately observed, recalled, and reported his experience. And he may have. But Samuels is a scientific expert on tiny fungi that grow on wood. His expertise is botany, not identifying large primates in poor conditions. Anyone, degreed or not, can be mistaken.

## Footprints

Bigfoot tracks are the most recognizable evidence; of course, the animal's very name came from the size of the footprints it leaves behind. Unlike sightings, they are physical evidence: *something* (known animal, Bigfoot, or man) left the tracks. The real question is what the tracks are evidence of. In many cases, the answer is clear: they are evidence of hoaxing.

Contrary to many Bigfoot enthusiasts' claims, Bigfoot tracks are not particularly consistent and show a wide range of variation (Dennett 1996). Some tracks have toes that are aligned, others show splayed toes. Most alleged Bigfoot tracks have five toes, but some casts show creatures with two, three, four, or even six toes (see figure 1). Surely all these tracks can't come from the same unknown creature, or even species of creatures.

Not all prints found are footprints, though. In September 2000, a team of investigators from the Bigfoot Field Research Organization led an expedition near Mt. Adams in Washington state, finding the first Bigfoot "body print," which-if authentic-is arguably the most significant find in the past two decades. The Bigfoot, according to the team, apparently made the impression when it laid on its side at the edge of a muddy bank and reached over to grab some bait. This of course raises the question as to why the animal would make such an odd approach to the food, instead of simply walking over to it and taking it. As the log of the expedition reads, "One explanation is immediately apparent-the animal did not want to leave tracks. . . ." (BFRO 2000). This explanation fails on its own logic: If the Bigfoot (or whatever it was) was so concerned about not leaving traces of its presence, why did it then leave a huge fifteen-square-foot imprint in the mud for the team to find?[1]

## Recordings

### The Patterson Film

The most famous recording of an alleged Bigfoot is the short 16 mm film taken in 1967 by Roger Patterson and Bob Gimlin. Shot in Bluff Creek, California, it shows a Bigfoot striding through a clearing (see figure 2). In many ways the veracity of the Patterson film is crucial, because the casts made from those tracks are as close to a gold standard as one finds in cryptozoology. Many in the Bigfoot community are adamant that the film is not-and, more important- *cannot* be a hoax. The question of whether the film is in fact a hoax or not is still open, but the claim that the film *could not* have been faked is demonstrably false.

Grover Krantz, for example, admits that the size of the creature in the film is well within human limits, but argues that the chest width is impossibly large to be human. "I can confidently state that no man of that stature is built that broadly," he claims (Krantz 1992, 118). This assertion was examined by two anthropologists, David Daegling and Daniel Schmitt (1999), who cite anthropometric literature showing the "impossibly wide" chest is in fact within normal human variation. They also disprove claims that the Patterson creature walks in a manner impossible for a person to duplicate.

The film is suspect for a number of reasons. First, Patterson told people he was going out with the express purpose of capturing a Bigfoot on camera. In the intervening thirty-five years (and despite dramatic advances in technology and wide distribution of handheld camcorders), thousands of people have gone in search of Bigfoot and come back empty-handed (or with little but fuzzy photos). Second, a known Bigfoot track hoaxer claimed to have told Patterson exactly where to go to see the Bigfoot on that day (Dennett 1996). Third, Patterson made quite a profit from the film, including publicity for a book he had written on the subject and an organization he had started.

In his book *Bigfoot*, John Napier, an anatomist and anthropologist who served as the Smithsonian Institution's director of primate biology, devotes several pages to close analysis of the Patterson film (pp. 89-96; 215-220). He finds many problems with the film, including that the walk and size is consistent with a man's; the center of gravity seen in the subject is essentially that of a human; and the step length is inconsistent with the tracks allegedly taken from the site. Don Grieve, an anatomist specializing in human gait, came to the conclusion that the walk was essentially human in type and could be made by a modern man. Napier writes that "there is little doubt that the scientific evidence taken collectively points to a hoax of some kind."

Other films and photos of creatures supposed to be Bigfoot have appeared, perhaps best-known among them the Wild Creek photos allegedly purchased by Cliff Crook of Bigfoot Central from an anonymous park ranger (see figure 3).

*Bigfoot Voices*
One of the more interesting bits of "evidence" offered for the existence of Bigfoot is sound recordings of vocalizations. One company, Sierra Sounds, markets a CD called "The Bigfoot Recordings: The Edge of Discovery." Narrated by Jonathan Frakes (an actor who also narrated a special on the infamous "Alien Autopsy" hoax), the recording claims to have captured vocalizations among a Bigfoot family. The sounds are a series of guttural grunts, howls, and growls.

The Web site and liner notes offer testimonials by "expert" Nancy Logan. Logan, their "linguist," apparently has little or no actual training (or degree) in linguistics. Her self-described credentials include playing the flute, speaking several languages, and having "a Russian friend [who] thinks I'm Russian." Logan confidently asserts that the tapes are not faked, and that the vocal range is too broad to be made by a human. She suggests that the Bigfoot language shows signs of complexity, possibly including profanities: "On one spot of the tape, an airplane goes by and they seem to get very excited and not very happy about it. Maybe those are Sasquatch swear words."

Here's what Krantz writes about Bigfoot recordings: "One... tape was analyzed by some university sound specialists who determined that a human voice could not have made them; they required a much longer vocal tract. A sasquatch investigator later asked one of these experts if a human could imitate the sound characteristics by simply cupping his hands around his mouth. The answer was yes" (Krantz 1992, 134). As for other such recordings, Krantz has "listened to at least ten such tapes and find[s] no compelling reason to believe that any of them are what the recorders claimed them to be" (133).

## Somatic Samples

Hair and blood samples have been recovered from alleged Bigfoot encounters. As with all the other evidence, the results are remarkable for their inconclusiveness. When a definite conclusion has been reached, the samples have invariably turned out to have prosaic sources-"Bigfoot hair" turns out to be elk, bear, or cow hair, for example, or suspected "Bigfoot blood" is revealed to be transmission fluid. Even advances in genetic technology have proven fruitless. Contrary to popular belief, DNA cannot be derived from hair samples alone; the root (or some blood) must be available.

In his book *Big Footprints*, Grover Krantz (1992) discusses evidence for Bigfoot other than footprints, including hair, feces, skin scrapings, and blood: "The usual fate of these items is that they either receive no scientific study, or else the documentation of that study is either lost or unobtainable. In most cases where competent analyses have been made, the material turned out to be bogus or else no determination could be made" (125). He continues, "A large amount of what looks like hair has been recovered from several places in the Blue Mountains since 1987. Samples of this were examined by many supposed experts ranging from the FBI to barbers. Most of these called it human, the Redkin Company found significant differences from human hair, but the Japan Hair Medical Science Lab declared it a synthetic fiber. A scientist at [Washington State] University first called it synthetic, then looked more closely and decided it was real hair of an unknown type. . . . Final confirmation came when E.B. Winn, a pharmaceutical businessman from Switzerland, had a sample tested in Europe. The fiber was positively identified as artificial and its exact composition was determined: it is a prod- uct known commercially as Dynel, which is often used as imitation hair." In his analysis, Winn (1991) noted that another alleged Bigfoot sign found at the site, tree splintering, had also been faked.

## Hoaxes, the Gold Standard, and the Problem of Experts

Such hoaxes have permanently and irreparably contaminated Bigfoot research. Skeptics have long pointed this out, and many Bigfoot researchers freely admit that their field is rife with fraud. This highlights a basic problem underlying all Bigfoot research: the lack of a standard measure. For example, we know what a bear track looks like; if we find a track that we *suspect* was left by a bear, we can compare it to one we *know* was left by a bear. But there are no undisputed Bigfoot specimens by which to compare new evidence. New Bigfoot tracks that don't look like older samples are generally not taken as proof that one (or both) sets are fakes, but instead that the new tracks are simply from a different Bigfoot, or from a different species or family. This unscientific lack of falsifiability plagues other areas of Bigfoot research as well.

Bigfoot print hoaxing is a time-honored cottage industry. Dozens of people have admitted making Bigfoot prints. One man, Rant Mullens, revealed in 1982 that he and friends had carved giant Bigfoot tracks and used them to fake footprints as far back as 1930 (Dennett 1996). In modern times it is easier to get Bigfoot tracks. With the advent of the World Wide Web and online auctions, anyone in the world can buy a cast of an alleged Bigfoot print and presumably make tracks that would very closely match tracks accepted by some as authentic.

What we have, then, are new tracks, hairs, and other evidence being compared to *known* hoaxed tracks, hairs, etc. as well as *possibly* hoaxed tracks, hairs, etc. With sparse hard evidence to go on and no good standard by which to judge new evidence, it is little wonder that the field is in disarray and has trouble proving its theories. In one case, Krantz claimed as one of the gold standards of Bigfoot tracks a print that "passed all my criteria, published and private, that distinguishes sasquatch tracks from human tracks and from fakes" (Krantz 1992). He further agreed that it had all the signs of a living foot, and that no human foot could have made the imprint. Michael R. Dennett, investigating for the *Skeptical Inquirer*, tracked down the anonymous construction worker who supplied the Bigfoot print. The man admitted faking the tracks himself to see if Krantz could really detect a fake (Dennett 1994).

Krantz certainly isn't alone in his mistaken identifications. One of the biggest names in cryptozoology, Ivan Sanderson, was badly fooled by tracks he confidently proclaimed would be impossible to fake. In 1948 (and for a decade afterward), giant three-toed footprints were found along the beach in Clearwater, Florida. Sanderson, described as a man who "was extremely knowledgeable on many subjects, and had done more fieldwork than most zoologists do today"

(Greenwell 1988), spent two weeks at the site of the tracks investigating, analyzing the tracks, and consulting other experts. He concluded that the tracks were made by a fifteen-foot-tall penguin.

In 1988, prankster Tony Signorini admitted he and a friend had made the tracks with a pair of cast iron feet attached to high-top black sneakers. J. Richard Greenwell, discussing the case in *The ISC Newsletter* (Winter 1988), summed the case up this way: "The lesson to be learned within cryptozoology is, of course, fundamental. Despite careful, detailed analyses by zoologists and engineers, which provided detailed and sophisticated mechanical and anatomical conclusions supporting the hypothesis of a real animal, we now see that, not only was the entire episode a hoax, but that it was perpetrated by relatively amateur, good-natured pranksters, *not* knowledgeable experts attempting, through their expertise, to fool zoological authorities."

The experts, however are only partly to blame for their repeated and premature proclamations of the authenticity of Bigfoot evidence. After all, other areas of science are not fraught with such deception and hoaxing; in physics and biology, light waves and protozoa aren't trying to trick their observers.

Even when there is no intentional hoaxing, "experts" have been fooled. In March 1986, Anthony Wooldridge, an experienced hiker in the Himalayas, saw what he thought was a Yeti (Himalayan Bigfoot) standing in the snow near a ridge about 500 feet away. He described the figure as having a head that was "large and squarish," and the body "seemed to be covered with dark hair." It didn't move or make noise, but Wooldridge saw odd tracks in the snow that seemed to lead toward the figure. He took two photos of the creature, which were later analyzed and shown to be genuine and undoctored. Many in the Bigfoot community seized upon the Wooldridge photos as clear evidence of a Yeti, including John Napier. Many suggested that because of his hiking experience it was unlikely Wooldridge made a mistake. The next year researchers returned to the spot and found that Wooldridge had simply seen a rock outcropping that looked vertical from his position. Wooldridge admitted his misidentification (Wooldridge 1987).

## Smoke and Fire

Bigfoot researchers readily admit that many sightings are misidentifications of normal animals, while others are downright hoaxes. Diane Stocking, a curator for the BFRO, concedes that about 70 percent of sightings turn out to be hoaxes or mistakes (Jasper 2000); Loren Coleman puts the figure even higher, at at least 80 percent (Klosterman 1999). The remaining sightings, that small portion of reports that can't be explained away, intrigue researchers and keep the pursuit active. The issue is then essentially turned into the claim that "Where there's smoke there's fire."

But is that really true? Does the dictum genuinely hold that, given the mountains of claims and evidence, there *must* be some validity to the claims? I propose not; the evidence suggests that there are enough sources of error (bad data, flawed methodological assumptions, mistaken identifications, poor memory recall, hoaxing, etc.) that there does not *have* to be (nor is likely to be) a hidden creature lurking amid the unsubstantiated cases.

The claim also has several inherent assumptions, including the notion that the unsolved claims (or sightings) are qualitatively different from the solved ones. But paranormal research and cryptozoology are littered with cases that were deemed irrefutable evidence of the paranormal, only to fall apart upon further investigation or hoaxer confessions. There will always be cases in which there simply is not enough evidence to prove something one way or the other. To use an analogy borrowed from investigator Joe Nickell, just because a small percentage of homicides remain unsolved doesn't mean that we invoke a "homicide gremlin"-appearing out of thin air to take victims' lives-to explain the unsolved crimes. It is not that such cases are *unexplainable* using known science, just that not enough (naturalistic) information is available to make a final determination.

A lack of information (or negative evidence) cannot be used as positive evidence for a claim. To do so is to engage in the logical fallacy of arguing from ignorance: We don't know what left the tracks or what the witnesses saw, therefore it must have been Bigfoot. Many Bigfoot sightings report "something big, dark, and hairy." But Bigfoot is not the only (alleged) creature that matches that vague description.

## The Future for Bigfoot

Ultimately, the biggest problem with the argument for the existence of Bigfoot is that no bones or bodies have been discovered. This is really the 800-pound Bigfoot on the researchers' backs, and no matter how they explain away the lack of other types of evidence, the simple fact remains that, unlike nearly every other serious "scientific" pursuit, they can't point to a live or dead sample of what they're studying. If the Bigfoot creatures across the United States are really out there, then each passing day should be one day closer to their discovery. The story we're being asked to believe is that thousands of giant, hairy, mysterious creatures are constantly eluding capture and discovery and have for a century or more. At some point, a Bigfoot's luck must run out: one out of the thousands must wander onto a freeway and get killed by a car, or get shot by a hunter, or die of natural causes and be discovered by a hiker. Each passing week and month and year and decade that go by without definite proof of the existence of Bigfoot make its existence less and less likely.

On the other hand, if Bigfoot is instead a self-perpetuating phenomenon with no genuine creature at its core, the stories, sightings, and legends will likely continue unabated for centuries. In this case the believers will have all the evidence they need to keep searching-some of it provided by hoaxers, others perhaps by honest mistakes, all liberally basted with wishful thinking. Either way it's a fascinating topic. If Bigfoot exist, then the mystery will be solved; if they don't exist, the mystery will endure. So far it has endured for at least half a century.

## NOTE

The way in which the track was discovered raises questions as well. The expedition log gives an account of how "[Team member Richard] Noll notices an unusual impression in the transition mud at the edge of the wallow and suddenly figures out what caused it. [Team members] Fish and Randles note the shock on Noll's face and come over to have another look at what he's examining. The three observe and note the various parts of the impression, and the chunks of chewed apple core nearby. The base camp is alerted. Everyone comes to see the impression. All conclude the animal was laying on its side at the edge of the mud, reaching out over the soft mud to grab the fruit" (BFRO 2000). So what you have is a case where a group of people are looking for evidence of a Bigfoot. One observer believes he sees a pattern fitting what he's looking for in ambiguous stimuli (shapes in mud). Once the pattern is pointed out to others, they also agree that the pattern could match up to parts of a hominid form in a particular contortion. The rest of the group, who might never have decided on their own that the pattern fits a Bigfoot, then validate the initial observer's (possibly unwarranted) conclusion. This happens all the time, for example when a person recognizes a face or an image in clouds or stains or tortillas. As psychologists know, observers' expectations frequently color their interpretations.

## REFERENCES

**Originally published in *Skeptical Inquirer* 26, no. 2 (March/April 2002): 29–34.**

Baird, D. 1989. Sasquatch footprints: A proposed method of fabrication. *Cryptozoology* 8: 43-46.

Betts, J. 1996. Wanted: Dead or alive. *Fortean Times* 93: 34-35, December.

BFRO. 2000. Account of the expedition. Bigfoot Field Research Organization. Available at www.bfro.net.

Big Disagreement Afoot. 2000. Associated Press report on ABCnews.com.

Bord, J., and Colin Bord. 1982. *The Bigfoot Casebook*. Harrisburg (Pa.): Stackpole Books.

Coleman, L. 1996. Footage furore flares. *Fortean Times* 91, October.

———. 1998. Suits you, sir! *Fortean Times* 106, January.

Coleman, L., and P. Huyghe. 1999. *The Field Guide to Bigfoot, Yeti, and Other Mystery Primates Worldwide*. New York: Avon Books.

Daegling, D., and D. Schmitt. 1999. Bigfoot's screen test. *Skeptical Inquirer* 23(3), May/June: 20-25.

Dennett, M. 1989. Evidence for Bigfoot? An investigation of the Mill Creek 'Sasquatch Prints.' *Skeptical Inquirer* 13(3), Spring: 264-272.

————. 1994. Bigfoot evidence: Are these tracks real? *Skeptical Inquirer* 18(5), Fall: 498-508.

————. 1996. Bigfoot. In Stein, G. (ed.) *Encyclopedia of the Paranormal.* Buffalo, N.Y.: Prometheus.

————. 2001. Personal communication, May 1.

Fahrenbach, W.H. 1998. Re: Interim statement on the Blue Mountain / Ohio hair. Available at Bigfoot Field Researcher's Homepage, www.bfro.net.

Freeland, D., and W. Rowe. 1989. Alleged pore structure in Sasquatch (Bigfoot) footprints. *Skeptical Inquirer* 13(3), Spring: 273-276.

Green, J. 1968. *On the Track of the Sasquatch* Cheam Publishing Ltd. Agassiz, B.C.

————. 2000. Green says Skookum Cast may be proof. In BFRO press release.

Greenwell, J.R. 1988. Florida "Giant Penguin" hoax revealed. *The ISC Newsletter.* 7(4), Winter.

Hubbell. J.M. 1999. Bigfoot enthusiasts discredit film. Associated Press report, January 10.

Jasper, D. 2000. Bigfoot strikes again! *Weekly Planet* October 26-November 1.

Klosterman, C. 1999. Believing in Bigfoot. *Beacon Journal* (Akron, Ohio), March 24.

Krantz, G. 1992. *Big Footprints: A Scientific Inquiry Into the Reality of Sasquatch.* Boulder: Johnson Books.

Napier, J. 1973. *Bigfoot: The Yeti and Sasquatch in Myth and Reality.* New York: E.P. Dutton & Co.

Roesch, B. 2001. On the nature of cryptozoology and science. *Fortean Times* online, March.

Winn, E. 1991. Physical and morphological analysis of samples of fiber purported to be Sasquatch hair. *Cryptozoology* 10: 55-65.

Wooldridge, A.B. 1987. The Yeti: A rock after all? *Cryptozoology* 6: 135.

Zuefle, D. 1999. Tracking Bigfoot on the Internet. *Skeptical Inquirer* 23(3), May/June: 26-28.

# ABOUT THE CONTRIBUTORS

**Stephen Barrett** is a retired psychiatrist, vice-president of the National Council Against Health Fraud, editor of *Consumer Health Digest*, scientific advisor to the American Council on Science and Health, and a fellow of the Committee for Skeptical Inquiry (CSI), formerly known as the Committee for the Scientific Investigation of Claims of the Paranormal (CSICOP). Dr. Barrett operates Quackwatch.org—a website designed to combat health-related fraud. He received the 2001 Distinguished Service to Health Education Award. He is medical editor of Prometheus Books, has more than 2,000 articles, delivered more than 300 talks, written 50 books, and has made numerous national media appearances.

**Barry Beyerstein** is Professor of Psychology and a member of the Brain Behaviour Laboratory at Simon Fraser University in British Columbia. His Ph.D. is in Experimental and Biological Psychology from the University of California at Berkeley. Dr. Beyerstein serves as chair of the Society of B.C. Skeptics and he is a Fellow and a member of the Executive Council of CSI and is on the editorial board of *Skeptical Inquirer*. He was also an elected to the Council for Scientific Medicine, and is a contributing editor of *The Scientific Review of Alternative Medicine*.

**Susan Blackmore** is a visiting lecturer at the University of the West of England, Bristol. She has a degree in psychology and physiology from Oxford University and a Ph.D. in parapsychology from the University of Surrey. Her research on consciousness, memes, and anomalous experiences has been published in over sixty academic papers, and book chapters. *The Meme Machine* has been translated into twelve other languages.

**Geoffrey Dean** is a technical editor in Perth, Western Australia. A former astrologer, and now a Fellow of CSI for his distinguished record of scholarship and significant contributions to science and skepticism, he and his associates have been investigating astrological claims via tests, surveys, debates, and prize competitions since the 1970s.

**Bryan Farha** is Professor of Behavioral Studies in Education at Oklahoma City University. He received the Doctor of Education degree in Counseling Psychology from the University of Tulsa. A *Licensed Professional Counselor,*

Dr. Farha is also a scientific and technical consultant to the *Committee for the Scientific Investigation of Claims of the Paranormal*, and a consulting editor for *The Scientific Review of Mental Health Practice*. With numerous publications, his work in combating deception has been recognized by TIME.com, *The National Geographic Channel*, CNN, and A&E.

**Ray Hyman** is Professor Emeritus in Psychology at the University of Oregon. He earned a Ph.D. degree in psychology from Johns Hopkins University and taught at Harvard. Also a skilled magician, Hyman was a founding member of the Committee for Skeptical Inquiry (CSI). Co-recipient of the 2005 Robert P. Balles Prize in Critical Thinking, Hyman has consulted with the United States Department of Defense in critically analyzing psychic research.

**Joe Nickell** is senior research fellow for the *Committee for the Scientific Investigation of Claims of the Paranormal* and earned a Ph.D. degree from the University of Kentucky. Author of the "Investigative Files" column for Skeptical Inquirer, Dr. Nickell is the author of sixteen books, including *Inquest of the Shroud of Turin*, *Secrets of the Supernatural*, *Looking for a Miracle*, *Entities*, *Psychic Sleuths*, and *The UFO Invasion*.

**Benjamin Radford** is managing editor of *Skeptical Inquirer* and has written over one hundred articles on topics including urban legends, mass hysteria, and media criticism. Radford has appeared on The Discovery Channel, the Learning Channel, the National Geographic Channel, and MTV's Big Urban Myth Show.

**James Randi** is founder of the James Randi Educational Foundation (JREF)—an organization promoting critical thinking about paranormal claims. A professional magician, he is known to many as "The Amazing Randi," is the author of eleven books, and is perhaps the world's best-known psychic investigator and skeptic. His writings have been published in prestigious periodicals, and in 1996 he received the MacArthur award for exposing the deception of alleged psychics. Randi was named one of the "100 Best People in the World" by Esquire magazine—a distinction shared with such notables as Nelson Mandela and Stephen Hawking.

**Emily Rosa** was a sixth-grade student in Loveland, Colorado, when her multi-authored article on therapeutic touch was published.

**Linda Rosa** is a Registered Nurse and holds a B.S. degree in nursing. At the time of her publication, she was working at the National Council Against Health Fraud.

**Carl Sagan** (1934–1996) was the David Duncan Professor of Astronomy and Space Sciences and Director of the Laboratory for Planetary Studies at Cornell University. He played a leading role in the Mariner, Viking, and Voyager spacecraft expeditions to the planets, for which he received the NASA Medals for Exceptional Scientific Achievement. His research on the origin of life began in the 1950s and his book *Cosmos* was the best-selling science book ever published in the English language. Co-founder of The Planetary Society, he served as Distinguished Visiting Scientist at the Jet Propulsion Laboratory, California Institute of Technology. Dr. Sagan received the Pulitzer Prize, and the Oersted Medal for his contributions to science, literature, education, and preservation of the environment.

**Larry Sarner** was affiliated with the National Therapeutic Touch Study Group at the time of his publication.

**Michael Shermer** is founder of the Skeptics Society and *Skeptic* magazine. He earned an M.A. in experimental psychology and a Ph.D. in the history of science. Dr. Shermer is a contributing editor and monthly columnist for *Scientific American*, and host of the Skeptics Distinguished Lecture Series at Caltech. He is co-host and producer of the Fox Family television series, *Exploring the Unknown*. Shermer is the author of many books and appears on numerous national television programs to provide a skeptical perspective of extraordinary claims. He has also appeared in documentaries on A&E, Discovery, PBS, The History Channel, The Science Channel, and The Learning Channel.

CPSIA information can be obtained at www.ICGtesting.com
Printed in the USA
BVOW05s0447241114

376371BV00004B/245/P